Gas Purification

Gas Purification

Editor

Milind Kulkarni

Gas Purification

Edited by **Milind Kulkarni**

Printed in 2017

ISBN: 978-1-68117-383-2

Library of Congress Control Number: 2015936532

© 2016 by
SCITUS Academics LLC,
616, Corporate Way, Suite 2, 4766,
Valley Cottage, NY 10989

www.scitusacademics.com

Contents

Preface

Gases are purified for the purpose of further using the gases themselves or the impurities contained in them; industrial gases discharged into the atmosphere are purified to prevent pollution of the air by noxious substances. Until the second half of the 19th century the struggle against the harmful effect of discharging industrial gases into the atmosphere consisted only of banning or restricting the construction of certain enterprises; however, these measures were rendered ineffective by the growth of industry, transportation, and large cities. The problem of purification of industrial gases arose precisely because of rapid industrial development, the concentration of industries, and increased scales of production. In industrially developed countries, the saturation of areas with industries and transportation was such that local pollution of the atmosphere became universal and led to the pollution of the entire air basin, or at least a huge part of it.

Editor

Synergistic Effect of Brønsted Acid and Platinum on Purification of Automobile Exhaust Gases

Wei Fu[1], Xin-Hao Li[1], Hong-Liang Bao[2], Kai-Xue Wang[1], Xiao Wei[1], Yi-Yu Cai[1], and Jie-Sheng Chen[1]

[1]School of Chemistry and Chemical Engineering, Hirano Institute for Materials Innovation, Shanghai Jiao Tong University, Shanghai 200240, China

[2]Shanghai Synchrotron Radiation Facilities, Shanghai Institute of Applied Physics, Chinese Academy of Sciences, Shanghai 201204, China

ABSTRACT

The catalytic purification of automobile exhaust gases (CO, NO_x and hydrocarbons) is one of the most practiced conversion processes used to lower the emissions and to reduce the air pollution. Nevertheless, the good performance of exhaust gas purification catalysts often requires the high consumption of noble metals such as platinum. Here we report that the Brønsted acid sites on the external surface of a microporous silicoaluminophosphate (SAPO) act as a promoter for exhaust gas purification, effectively cutting the loading amount of platinum in the catalyst without sacrifice of performance. It is revealed that in the Pt-loaded SAPO-CHA catalyst, there exists a remarkable synergistic effect between the Brønsted acid sites and the Pt nanoparticles, the former helping to adsorb and activate the hydrocarbon molecules for NO reduction during the catalytic process. The thermal stability of SAPO-CHA also makes the composite catalyst stable and reusable without activity decay.

INTRODUCTION

The simultaneous catalytic removal of major pollutants CO, NO_x and hydrocarbons (HCs) from the exhaust gases of a gasoline engine is an important conversion process to reduce the emissions of toxic gases from automobiles. Platinum-containing catalysts have attracted increasing attention in the past decades because of their superior activity as a catalyst not only for redox reactions but also for purification of automobile exhaust gases[1, 2, 3, 4, 5]. Many researches have been focused on the development of Pt-loaded catalysts that are desired to effectively eliminate the emission of harmful gases under practical operation conditions[4]. Despite great efforts to date with significant successes, the good performance of catalysts usually leads to high consumption of expensive noble metals, and recovery of noble metals from spent catalysts is required. In general, the performance of a catalyst is also heavily dependent on the physical and chemical properties of its support, which should possess not only a considerable surface area to sustain a uniform dispersion of noble metals, but also strong interaction with the metals to promote the redox reactions[6]. In this context, the selection of suitable supports is very critical in determining the performance and the cost of Pt-containing catalysts for purifying emission gases.

To optimize catalyst supports for stabilizing Pt nanoparticles, materials that allow access to activating substrates and/or elevating the activity of Pt nanoparticles are preferred. Microporous crystalline silicoaluminophosphates (SAPOs) are an important class of molecular sieve materials which have been widely used in adsorption, separation

Synergistic Effect of Brønsted Acid and Platinum on Purification of...

3

and catalysis. Among the SAPOs, chabazite-type SAPO (SAPO-CHA)[7] with small pores and fairly strong Brønsted acidity has shown excellent catalytic activity and shape selectivity in methanol-to-olefin (MTO) conversion[8, 9,10] and HC transformation[11, 12, 13]. It is thus rather promising to integrate the catalytic activity of SAPO-CHA with rich Brønsted acid sites and Pt nanoparticles for efficient removal of automobile exhaust gases, which also contain HCs, under even mild conditions. Although previous studies have used SAPO-CHA containing metal cations for removal of NO_x through a selective-catalytic-reduction (SCR) pathway[14, 15, 16], the exploitation of rich Brønsted acid sites in SAPO-CHA support for activating noble metal nanoparticles to simultaneously purify the major pollutants CO, NO_x and HC of a gasoline engine emission has not been described yet. In this report, we demonstrate the remarkable promotion effect of Brønsted acid sites of SAPO-CHA on the catalytic performance of supported Pt catalyst for purification of automobile exhaust gases. Accordingly, the catalyst cost may be reduced markedly by decreasing the Pt-loading amount on the SAPO-CHA material without sacrifice of catalytic performance.

RESULTS

Preparations of Catalysts

We prepared the microporous SAPO-CHA with a crystal size of 1.02 μm and an outer surface area of 32 m^2 g^{-1} through a typical hydrothermal route (see Methods). An aluminophosphate analogue (AlPO-CHA) with similar CHA framework structure was also prepared as Brønsted acid-free reference material[17, 18]. In SAPO-CHA, a part of the PO_4 tetrahedra are replaced by SiO_4 units, resulting in negative charges which may be balanced by extra-framework protons, that is, Brønsted acid sites discussed in this work. Pt nanoparticles were loaded on the surface of AlPO-CHA and SAPO-CHA through an impregnation approach, and the resulting materials are designated as Pt/AlPO-CHA and Pt/SAPO-CHA, respectively. For comparison, a Pt-loaded alumina material (Pt/Al_2O_3) was also prepared similarly and used as a reference. Pt nanoparticles were mainly dispersed on the external surface of the molecular sieves as the size of Pt nanoparticles observed in the TEM image is much larger that the pore size of the molecular sieves.

Distribution of Brønsted Acid Sites

The distribution of Brønsted acid sites was initially investigated by temperature-programmed desorption (TPD) measurements (Fig. 1a), where NH_3 and pyridine were used as the probe molecules. The NH_3–TPD profiles with two strong desorption peaks at temperatures of 680 and 850 K reveal a rich amount of Brønsted acid sites on the surface of SAPO-CHA based materials and also the Brønsted-acid-free nature of AlPO-CHA materials (Figure 1a). This observation has been doubly confirmed by the Fourier transform infrared (FTIR) spectra of dehydrated SAPO-CHA with strong absorptions at 3630 and 3600 cm^{-1} (Fig. S3). Pyridine molecules with a large dynamic diameter (0.65 nm) cannot diffuse into the micropores (window size 0.43 nm) of SAPO-CHA, and were thus used as probe molecules to characterize the acidity of the external surfaces of SAPO-CHA. The strong signal at 1540 cm^{-1}, arising from protonated pyridinium cations, in the FTIR spectrum (Fig. 1b) of pyridine-adsorbed samples indicates a substantial amount of Brønsted acid sites on the external surface of SAPO-CHA, whilst such a signal disappeared in the spectra of Brønsted-acid-free AlPO-CHA.

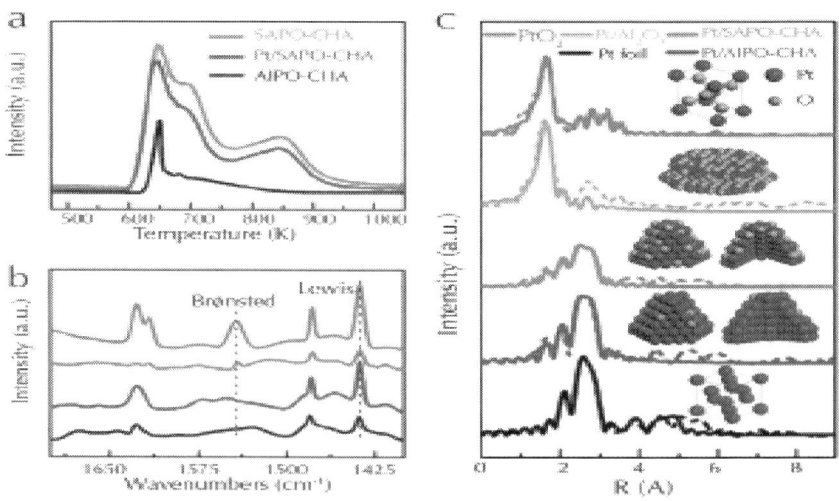

Figure 1: General characterization of Pt/AlPO-CHA, Pt/SAPO-CHA and Pt/Al_2O_3. (a) NH_3-TPD profiles. There are rich Brønsted acid sites on the internal and external surfaces of the as-prepared Pt/SAPO-CHA, while no such sites exist in the Pt/AlPO-CHA material. (b) FTIR spectra of pyridine-adsorbed

Synergistic Effect of Brønsted Acid and Platinum on Purification of...

5

SAPO-CHA outgased at 573 K (red) and 373 K (green) and AlPO-CHA out-gased at 573 K (black) and 373 K (blue). (c) Fourier-transformed EXAFS spectra (solid lines) along with theoretical fits (dashed lines), which indicate that the Pt species on SAPO-CHA and AlPO-CHA are mainly metal nanoparticles, whereas on Al_2O_3 they are mainly Pt oxide. a.u. = arbitrary units.

Chemical State of the Pt on the Surface of Various Supports

The Pt nanoparticles are believed to be highly active in many important catalytic reactions as compared with their oxide PtO_2[19, 20]. For conventional Pt-containing catalysts (e.g. Pt/Al_2O_3), a pre-activation process to reduce the PtO_2 component to metallic Pt has to be conducted at an elevated temperature before practical use and/or reuse, resulting in additional cost due to the consumption of H_2 gas and electric energy. The stability of the metallic Pt should also be seriously considered in designing new sustainable catalysts. Both the X-ray absorption near-edge structure (XANES) spectra and Fourier transformed extended X-ray absorption fine structure (EXAFS) spectra (Figure 1c) of the Pt species dispersed on SAPO-CHA and AlPO-CHA are similar to those of Pt foil, rather speaking for the formation of metallic Pt nanoparticles. As described previously, the Pt species on Al_2O_3 exist mainly as Pt oxide[21, 22]. The representative schematic illustrations for the Pt species dispersed on different supports are shown in the inset of Fig. 1c. It can be envisaged that the nature of the support gives a significant impact on the state of Pt species generated on the support. No obvious reduction peaks were observed for Pt/SAPO-CHA and Pt/AlPO-CHA materials in the temperature-programmed reduction (TPR) experiments (Fig. S5), and this is to say that the high-temperature activation process in H_2 is not essential for Pt/SAPO-CHA catalyst. It is thus quite convenient for Pt/SAPO-CHA to be directly used in purification of automobile exhaust gases.

The Catalytic Performance

The catalytic activities of the Pt-loaded materials were tested in the purification of the exhaust gases of a gasoline engine. The conversions of exhaust gases (NO, CO and C_3H_6) over Pt/SAPO-CHA and control catalysts were monitored at gradually increased reaction temperatures. Previously, C_3H_6 has been widely applied as a typical model of HCs to evaluate the catalytic performance of catalysts in purification of automobile exhaust gases, and it was thus used in this work accordingly. As shown in Figure 2, the conversions of C_3H_6, NO and CO over Pt/SAPO-CHA are all higher than those of Pt/AlPO-CHA and Pt/Al$_2$O$_3$ with the same Pt-loading content (0.5 wt%). The temperature at a conversion of 50% of C_3H_6 over Pt/SAPO-CHA is 549 K, nearly 100 and 50 K lower than that over Pt/Al$_2$O$_3$ and Pt/AlPO-CHA respectively (Fig. 2), while the temperature at a conversion of 90% of C_3H_6 is about 578 K for Pt/SAPO-CHA, also much lower than those of Pt/AlPO-CHA (630 K) and Pt/Al$_2$O$_3$ (655 K). Moreover, bare SAPO-CHA as the catalyst gave no conversion of C_3H_6, NO and CO, suggesting that only Pt particles acted as the active cites for such a catalytic reaction. The turnover frequencies (TOF) over Pt/SAPO-CHA were calculated to be 0.014, 0.019 and 0.217 s^{-1} respectively for C_3H_6, NO and CO at a temperature of 575 K. Furthermore, products of NO conversion have been identified through gas chromatograph (GC) and gas-phase FTIR measurements. Typical IR bands of N_2O gas at 2212 and 2235 cm^{-1} were detected in the spectra of products catalyzed over Pt/AlPO-CHA and Pt/Al$_2$O$_3$, showing the moderate selectivity of these catalysts. The fact that the peaks of N_2O gas were not present in the spectra of purified gases over Pt/SAPO-CHA reveals the high selectivity of this catalyst in the purification of exhausting gases, again speaking for the advantage of SAPO-CHA as the catalyst support here.

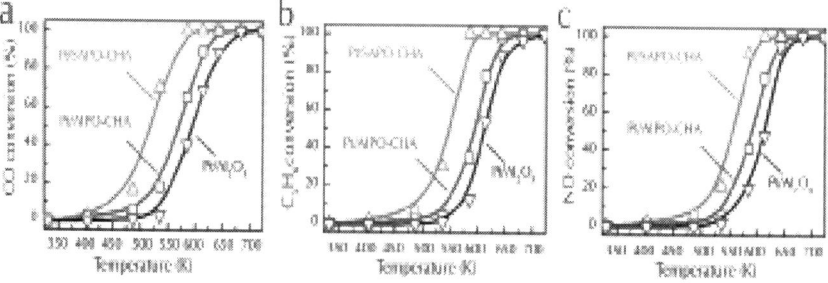

Figure 2: Catalytic performances of Pt/AlPO-CHA, Pt/SAPO-CHA and Pt/Al$_2$O$_3$ with the same Pt content of 0.5 wt%. (a) Conversion of CO, (b) conversion of C_3H_6, and (c) conversion of NO.

Synergistic Effect of Brønsted Acid and Platinum on Purification of...

7

The content of Pt loaded also affects the catalytic performance of the resulting catalysts to a considerable extent (Fig. S7). The activity of the catalysts increased with increasing amount of Pt components. It is noteworthy that the catalytic performance of Pt/SAPO-CHA with a Pt loading as low as 0.25 wt% is comparable with that of Pt/Al$_2$O$_3$ with a Pt loading as high as 1.0 wt%. As an excellent catalyst support, SAPO-CHA can significantly reduce the amount of the noble metal Pt and thus the cost of the catalyst.

DISCUSSION

To find out how the Brønsted acid affects the catalytic activity, it is essential to understand the nature of the acid sites in the SAPO-CHA molecular sieve. Individual Brønsted acid site is a hydroxyl proton located at bridging oxygen between tetrahedrally coordinated silicon and aluminum atoms (Figure 3b). Previous studies indicated that adsorption of olefins (e.g. C$_3$H$_6$ here) on Brønsted acid sites tend to be transformed to alkoxy species with a carbeniumion-like feature, which can undergo further reactions to become other derivative products in various task-specific catalytic reactions[28, 29, 30]. We thus calculated the structure of the SAPO-CHA using relativistic density functional theory (DFT) to elucidate the function of Brønsted acid in the catalytic transformation of C$_3$H$_6$. The simulation results suggest a very high electron density around the Brønsted acid sites (Fig. 3b), which facilitates the adsorption and activation of C$_3$H$_6$ molecules for conversion into covalent alkoxy species in the catalytic reactions[31]. Moreover, the high electron density of the Pt nanoparticles will also be enhanced once Pt nanoparticles form a contact at the electron-rich surface of SAPO-CHA, promising a possibility to improve the catalytic activity of Pt nanoparticles. The facts that the bare SAPO-CHA with rich Brønsted acid sites offered no obvious activity for purifying exhaust gases and gave much better catalytic performance as catalyst support than that of Brønsted-acid-site-free AlPO-CHA reveals an obvious synergistic effect between the Brønsted acid sites and the Pt nanoparticles. Such a synergistic effect facilitates conversion of C$_3$H$_6$ and NO gases via simultaneously lowering the energy barrier of the oxidation reaction of HCs and enhancing the activity of Pt nanoparticles.

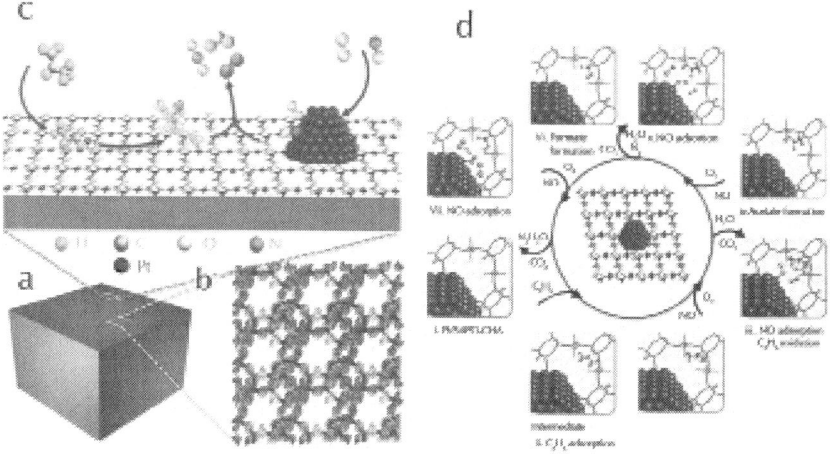

Figure 3: Schematic representation for the effect of Brønsted acid sites on catalytic reactions of exhaust gases. (a) The particle of the SAPO-CHA crystal. (b) The simulated STM image of SAPO-CHA, showing high electron density (red) around the Brønsted acid site. (c) The activation of a propene molecule facilitated by the Brønsted acid site. (d) The proposed synergistic mechanism between the surface Brønsted acid sites and the Pt nanoparticles on the external surface of SAPO-CHA.

On the basis of our experimental and calculation results in combination with the observations reported in the literature[32, 33, 34, 35], a catalytic cycle for the exhaust gas purification reaction over the Pt/SAPO-CHA catalyst is proposed and depicted in Fig. 3d. First, the C_3H_6 molecules are attracted by the Brønsted acid sites (step i), followed by conversion into covalent alkoxy species (step ii). These species are oxidized by O_2 over adjacent Pt nanoparticle, leading to the formation of acetate, H_2O, and CO_2 (step iii, iv). The acetate further react with NO to form nitrogen-containing organic intermediates (step v), which are further converted via formate (step vi) and HNCO species and finally into N_2 and CO_2 by following the reaction of $NCO^- + NO = N_2 + CO_2$ (step vii).

Reusability of the Pt/SAPO-CHA catalysts is another important aspect for practical application. To enhance the stability and thus keep the activity of Pt-nanoparticle based catalysts, the morphology and composition of the as-formed Pt nanoparticles should be maintained during catalytic reactions. The advantage of SAPO-CHA over the conventional γ-Al_2O_3 as a catalyst support to prevent the metallic Pt from oxidation has been demonstrated earlier in this communication, and is also confirmed by HRTEM analysis. It is found that the SAPO-CHA as well as AlPO-CHA with abundant PO_4 tetrahedra in the framework and the external surface can act as efficient stabilizer to improve the homogeneous dispersion of Pt nanoparticles and keep them from agglomeration during reactions (Fig. S8).

The thermal and chemical stability of catalyst supports is equally important to ensure their reusability. The structure and crystallinity of SAPO-CHA remains unchanged after reaction as confirmed by the FTIR (Fig. S9) and XRD analysis. As conventional catalysts for hydrogen transformation, zeolitic catalysts, including SAPO-CHA here, generally suffer from deactivation induced by the formation of coke inside the micropores. As expected, the thermogravimetric analysis (TGA) result reveals formation of coke (20 wt%) in the used Pt/SAPO-CHA catalyst Nevertheless, the conversions of CO, NO and C_3H_6 over Pt/SAPO-CHA at a space velocity of 34,000 mL gcat^{-1} h^{-1}, do not show any obvious decay over a 1,000-minute run at 620 K, suggesting that the coke formed during the catalytic reaction has no influence on activity of the Pt/SAPO-CHA catalyst. The NH_3-TPD and pyridine-TPD results suggest that coke forms only inside the micropores of SAPO-CHA and the amount of the Brønsted acid sites on the external surface is not varied. This observation well explains why the formation of coke leads to negligible deactivation as the Pt nanoparticles are mostly distributed on the external surface of the Pt/SAPO-CHA catalyst. On the other hand, it is exactly the Pt nanoparticles that catalyzes the complete transformation of propene to CO_2 and H_2O with concomitant NO reduction, and prevents the formation of coke on the external surface of the SAPO-CHA material.

METHODS

Materials

All the reagents for synthesis and preparation were of analytic grade and used as received without further purification. Aluminum hydroxide and fumed silica were purchased from Aldrich. Phosphoric acid (85 wt%), hydrofluoric acid (40 wt%), cyclohexylamine and morpholine were all purchased from Shanghai Chemical Reagent Factory. γ-alumina (Al_2O_3 with a BET surface area of 200 m^2 g^{-1}) was provided by Beijing Chemical Factory.

Preparation of SAPO-CHA and Alpo-CHA

The SAPO-CHA precursor was synthesized by a hydrothermal method from the gel with a molar composition of 1.0cyclohexylamine: $0.5SiO_2$: $1.0Al_2O_3$: $0.9P_2O_5$: $55H_2O$ following the procedure reported by Ashtekar et al[36]. Typically, 3.7 g of pseudoboehmite was first dispersed into 20 g of distilled water with stirring, followed by the addition of 0.75 g of fumed silica. After stirred at room temperature for one hour, 4.5 g of H_3PO_4 and then 3.6 g of cyclohexylamine were added slowly to the above mixture under vigorous stirring. After three hours, the resulting mixture was loaded to a Teflon-lined stainless steel autoclave and heated at 453 K for 10 days. The crystalline product was filtered off, washed with deionized water and dried at 333 K overnight. The as-synthesized material was calcined at 873 K in air for 24 hours to remove the template. The molar ratio Si: Al: P of the obtained SAPO-CHA was estimated to be 1:4.3:3.5 from the XRF results.

For comparison, microporous AlPO-CHA without Brønsted acidity was also prepared hydrothermally from the gel with a molar composition of 1.0 morpholine: $1.0 Al_2O_3$: $1.0 P_2O_5$: $60 H_2O$: 0.02 HF, following the procedure reported in the literature[37].

Loading of Pt

Pt was loaded onto SAPO-CHA, AlPO-CHA, and $γ-Al_2O_3$ by a modified wet impregnation method, leading to the formation of Pt/SAPO-CHA, Pt/AlPO-CHA, and Pt/Al_2O_3, respectively. Prior to the impregnation, SAPO-CHA and AlPO-CHA were calcined at 873 K in air for 24 hours to decompose the organic template within the micropores, while $γ-Al_2O_3$ was dried at 383 K for five hours to remove the physically adsorbed H_2O molecules and other species. A certain amount of H_2PtCl_6 was dissolved into 10 mL of ethanol. The calcined SAPO-CHA, AlPO-CHA, and $γ-Al_2O_3$ powders were then dispersed into the above ethanolic solution, respectively. These mixtures were kept at 333 K for two hours with vigorous stirring. Finally, Pt loaded samples, Pt/SAPO-

CHA, Pt/AlPO-CHA, and Pt/Al$_2$O$_3$, were obtained by removal of the ethanol solvent with a rotary evaporator at 333 K. After dried at 383 K for 12 hours, the Pt loaded samples were sieved to 20–40 meshes. For comparison, a Pt/Al$_2$O$_3$ sample was also prepared through aging the Pt-impregnated Pt/Al$_2$O$_3$ at approximately 1173 K for five hours, and the corresponding sample was designated Pt/Al$_2$O$_3$-aging.

General Characterization

The powder X-ray diffraction (XRD) patterns were recorded on a Rigaku Dmax-2200 diffractometer (Rigaku, Japan) with Cu Kα radiation (λ = 1.5418 Å). The distribution and crystal lattice of Pt were examined with a transmission electron microscope (JEM-2100F, JEOL, Japan), operating at 200 kV. The morphology of the samples was observed with a scanning electron microscope (SEM) (JEOL JSM-6700F, Japan), operated at 5 kV. Nitrogen adsorption/desorption analyses were performed on a Micromeritics ASAP 2010 M + C nitrogen adsorption instrument (Micrometritices Inc., USA) at 77 K. The contents of Pt loaded in the samples were determined by inductively coupled plasma spectroscopy (Thermo Electron Corporaton). Approximately 0.5 wt% Pt was loaded in the Pt/SAPO-CHA, Pt/AlPO-CHA, and Pt/Al$_2$O$_3$ samples.

Hydrogen temperature programmed reduction (H$_2$-TPR) of the samples was performed in a quartz tube reactor equipped with a thermal conductivity detector (TCD). 200 mg of sample was pretreated under He flow (50 mL min^{-1}) at 673 K for 30 min to remove the adsorbed water. After cooled to 323 K, the sample was exposed to a flow of 5% H$_2$/He (50 mL min^{-1}) and the temperature was raised to 1073 K at a rate of 10 K min^{-1}. The TPD experiments of adsorbed pyridine and NH$_3$ were performed in a quartz tube reactor equipped with a TCD. After saturated with ammonia or pyridine, the samples were purged with pure He at 373 K for two hours. Then, the samples were heated to 973 K at ramp rate of 10 K min^{-1} under helium flow (50 mL min^{-1}). The FTIR spectra were recorded on a Bruker IFS 66 v/S FTIR spectrometer equipped with a deuterated triglycine sulfate (DTGS) detector.

The Pt L_3-edge X-ray absorption spectra (XAS) were obtained on the BL14W1 beamline at the Shanghai Synchrotron Radiation Facility (SSRF), operated at 3.5 GeV with injection currents of 140–210 mA. A Si (111) double-crystal monochromator was used to reduce harmonic component of the monochrome beam. All the samples were measured in fluorescence mode. The powder of a sample was pressed into a self-supporting disk, which was then sealed with Kapton membrane and subjected to XAS measurement. The IFEFFIT software was used to calibrate the energy scale, to correct the background signal and to normalize the intensity. Reliable parameters for the Z (Pt, Pt) and Z (Pt, O) contributions were determined by multiple-shell fitting in r space with application of k^3 and k^1 weightings in the Fourier transformations.

DFT Calculations of the Charge Density around Proton H⁺ In SAPO-CHA

Calculations were carried out using the CASTEP periodic density functional theory (DFT) package. The exchange-correlation energy and potential were described self-consistently within the generalized gradient approximation (GGA-PW91). The self-consistent PW91 density was determined by iterative diagonalization of the Kohn−Sham Hamiltonian. The wave functions were expanded with the plane wave, and the ultra-soft pseudo-potential method was used to reduce the number of plane waves. Plane waves were used as a basis set with an energy cutoff of 340 eV. The electron density was approximated using a multipolar expansion up to hexadecapole. Typical overbinding associated with local density functionals was rectified through the use of the gradient-corrected Perdew-Burke-Ernzerhof (PBE) functional. The Brillouin-zone integration was performed using a $2 \times 2 \times 2$ Monkhorst–Pack (MP) grid. The STM images were simulated using the Tersoff–Hamann approach with a bias voltage of 2 V.

Evaluation of the Catalytic Property

The catalytic performance of the samples for purifying emission gases was evaluated in a fixed-bed reactor. Approximately 250 mg of the sample was loaded into a tube quartz reactor, and then was reduced *in situ* with the simulated feed gas, containing 0.82 vol% O_2, 0.12 vol% NO, 1.10 vol% CO, 0.08 vol% C_3H_6 and balance He at 673 K for two hours. After cooling to 303 K, the feed gas was allowed to pass through the reactor at a flow rate of 180 mL min⁻¹ (corresponding to a space velocity of 43,200 mL h⁻¹ gcat⁻¹). The effluent gas composition was analyzed on-line by two gas chromatographs (Shimadzu, GC-2014) equipped with a 5A column coupled with a thermal conductivity detector, and an Rtx-1 column coupled with a flame-ionization detector. Concentrations of NO_x were determined by FTIR analyzer with Bruker IFS 66 v/S FTIR spectrometer, equipped with a quartz sample holder with CaF_4 windows.

AUTHOR CONTRIBUTIONS

W.F. and J.S.C. developed the idea and designed the experiments. W.F. performed the sample fabrication, measurements and data analysis. W.F., H.L.B., X.H.L., K.X.W., X.W., Y.Y.C. and J.S.C. analyzed the data, and discussed the results. W.F., X.H.L., K.X.W. and J.S.C. co-wrote the paper. J.S.C. planned and supervised the project.

ACKNOWLEDGEMENTS

This work was financially supported by the National Basic Research Program of China (2013CB934102, 2011CB808703), the National Natural Science Foundation, and the Shoei Chemical Inc. We thank Shanghai Synchrotron Radiation Facility for the X-ray absorption measurement.

REFERENCES

1. Labinger, J. A. & Bercaw, J. E. Understanding and exploiting C-H bond activation. *Nature* 417, 507–514 (2002).

2. Greeley, J. *et al.* Alloys of platinum and early transition metals as oxygen reduction electrocatalysts. *Nat. Chem.* 1, 552–556 (2009).

3. Strmcnik, D. *et al.* Enhanced electrocatalysis of the oxygen reduction reaction based on patterning of platinum surfaces with cyanide. *Nat. Chem.* 2, 880–885 (2010).

4. Kim, C. H., Qi, G., Dahlberg, K. & Li, W. Strontium-doped perovskites rival platinum catalysts for treating NO_x in simulated diesel exhaust. *Science* 327, 1624–1627 (2010).

5.   Yamamoto, K. *et al.* Size-specific catalytic activity of platinum clusters enhances oxygen reduction reactions. *Nat. Chem.* 1, 397–402 (2009).

6.   Granger, P. & Parvulescu, V. I. Catalytic NO_x abatement systems for mobile sources : from three-way to lean burn after-treatment technologies. *Chem. Rev.* 111, 3155–3207 (2011).

7.   Lok, B. M. *et al.* Silicoaluminophosphate molecular sieves: another new class of microporous crystalline inorganic solids. *J. Am. Chem. Soc.* 106, 6092–6093 (1984).

8.   Marcus, D. M. *et al.* Experimental evidence from H/D exchange studies for the failure of direct C-C coupling mechanisms in the methanol-to-olefin process catalyzed by HSAPO-34. *Angew. Chem., Int. Ed.* 118, 3205–3208 (2006).

9.   Haw, J. F. *et al.* The mechanism of methanol to hydrocarbon catalysis. *Acc. Chem. Res.* 36, 317–326 (2003).

10. Wang, W., Buchholz, A., Seiler, M. & Hunger, M. Evidence for an initiation of the methanol-to-olefin process by reactive surface methoxy groups on acidic zeolite catalysts. *J. Am. Chem. Soc.* 125, 15260–15267 (2003).

11. Boronat, M., Viruela, P. & Corma, A. Reaction intermediates in acid catalysis by zeolites: prediction of the relative tendency to form alkoxides or carbocations as a function of hydrocarbon nature and active site structure. *J. Am. Chem. Soc.* 126, 3300–3309 (2004).

12. Tuma, C. & Sauer, J. Protonated isobutene in zeolites: *tert*-butyl cation or alkoxide. *Angew. Chem., Int. Ed.* 44, 4769–4771 (2005).

13. Rozanska, X. *et al.* A periodic DFT study of isobutene chemisorption in proton-exchanged zeolites: dependence of reactivity on the zeolite framework structure. *J. Phys. Chem. B* 107, 1309–1315 (2003).

14. Wang, J. *et al.* The influence of silicon on the catalytic properties of Cu/SAPO-34 for NO_x reduction by ammonia-SCR. *Appl. Catal. B-Environ.* 27, 137–147 (2012).

15. Martinez-Franco, R. *et al.* Rational direct synthesis methodology of very active and hydrothermally stable Cu-SAPO-34 molecular sieves for the SCR of NO_x. *Appl. Catal. B-Environ.* 127, 273–280 (2012).

16. Fickel, D. W. *et al.* The ammonia selective catalytic reduction activity of copper-exchanged small-pore zeolites. *Appl. Catal. B-Environ.* 102, 441–448 (2011).

17. Li, S., Falconer, J. L. & Noble, R. D. Improved SAPO-34 membranes for CO_2/CH_4 separations. *Adv. Mater.* 18, 2601–2603 (2006).

18. Zhang, L. *et al.* Investigations of formation of molecular sieve SAPO-34. *J. Phys. Chem.* C115, 22309–22319 (2011).

19. Zhai, Y. *et al.* Alkali-stabilized $Pt-OH_x$ species catalyze low-temperature water-gas shift reactions. *Science* 329, 1633–1636 (2010).

20. Cortes, J. M. *et al.* Comparative study of Pt-based catalysts on different supports in the low-temperature de-NO_x-SCR with propene. *Appl. Catal. B-Environ.* 30, 399–408 (2001).

21. Borgna, A. *et al.* Sintering of Pt/Al_2O_3 reforming catalysts: EXAFS study of the behavior of metal particles under oxidizing atmosphere. *Catal. Lett.* 13, 175–188 (1992).

22. Cho, I. H. *et al.* Investigation of $Pt/ -Al_2O_3$ catalysts prepared by sol-gel method: XAFS and ethane hydrogenolysis. *J. Catal.* 173, 295–303 (1998).

23. Wang, Q. Y. *et al.* The effect of La doping on the structure of $Ce_{0.2}Zr_{0.8}O_2$ and the catalytic performance of its supported Pd-only three-way catalyst. *Appl. Catal. B* 101, 150–159(2010).

24. He, H. *et al.* Pd-, Pt-, and Rh-loaded $Ce_{0.6}Zr_{0.35}Y_{0.05}O_2$ three-way catalysts: an investigation on performance and Redox properties. *J. Catal.* 206, 1–13 (2002).

25. Guilhaume, N. & Primet, M. Three-way catalytic activity and oxygen storage capacity of perovskite $LaMn_{0.976}Rh_{0.024}O_{3 +}$. *J. Catal.* 165, 197–204 (1997).

26. Zhou, K. B. *et al.* Pd-containing perovskite-type oxides used for three-way catalysts. *J. Mol. Catal. A-Chem.* 189, 225–232 (2002).

27. Papavasiliou, A. *et al.* An investigation of the role of Zr and La dopants into $Ce_{1-x-y}Zr_x la_y O$ enriched $-Al_2O_3$ TWC washcoats. *Appl. Catal. A* 383, 73–84 (2010).

28. Farneth, W. E. & Gorte, R. J. Methods for characterizing zeolite acidity. *Chem. Rev.* 95,615–635 (1995).

29. Long, R. Q. & Yang, R. T. In situ FT-IR study of Rh-Al-MCM-41 catalyst for the selective catalytic reduction of nitric oxide with

propylene in the presence of excess oxygen. *J. Phys. Chem. B* 103, 2232–2238 (1999).

30. Santen, R. A. & Kramer, G. J. Reactivity theory of zeolitic broensted acidic sites. *Chem. Rev.* 95, 637–660 (1995).

31. Wang, W. & Hunger, M. Reactivity of surface alkoxy species on acidic zeolite catalysts.*Acc. Chem. Res.* 41, 895–904 (2008).

32. Xin, M., Hwang, I. C. & Woo, S. I. *In situ* FTIR study of the selective catalytic reduction of NO on Pt/ZSM-5. *Catal. Today.* 38, 187–192 (1997).

33. perez-Ramirez, J. *et al.* Characterization and performance of Pt-USY in the SCR of NO$_x$ with hydrocarbons under lean-burn conditions. *Appl. Catal. B* 29, 285–298 (2001).

34. Bamwenda, G. R. *et al.* Selective reduction of nitric oxide with propene over platinum-group based catalysts: studies of surface species and catalytic activity. *Appl. Catal. B* 6, 311–323 (1995).

35. Captain, D. K. & Amiridis, M. NO reduction by propylene over Pt/SiO$_2$: An *in situ* FTIR study. *J. Catal.* 194, 222–232 (2000).

36. Ashtekar, S., Chilukuri, S. V. V. & Chakrabarty, D. K. Small-pore molecular sieves SAPO-34 and SAPO-44 with chabazite structure: A study of silicon incorporation. *J. Phys. Chem.* 98, 4878–4883 (1994).

37. Marchese, L. *et al.* AlPO-34 and SAPO-34 synthesized by using morpholine as templating agent. FTIR and FT-Raman studies of the host-guest and guest-guest interactions within the zeolitic framework. *Micropor. Mesopor. Mater.* 30, 145–153 (1999).

Gel Permeation Chromatography Purification and Gas Chromatography-Mass Spectrometry Detection of Multi-Pesticide Residues in Traditional Chinese Medicine

Wan-E Zhuang, Zhen-Bin Gong

State Key Laboratory of Marine Environmental Science, College of Oceanography & Environmental Science, Xiamen University, Xiamen, China

ABSTRACT

The measurement of 23 organochlorine, organophosphorus, and pyrethroid pesticides in typical traditional Chinese medicine (TCM), flos lonicerae, was made using gel permeation chromatography (GPC) purification and gas chromatography-mass spectrometry (GC-MS) detection. The pesticides were extracted with ultrasonic device and 5.0 mL mixture of ethyl acetate and cyclohexane (1:1, v/v). Coextractants from sample matrices which may have interfere to the qualitative and quantitative analysis, such as pigments, were removed using GPC purification. Simultaneous full scan and selective ion monitor (scan/SIM) mode for GC-MS was used for qualitative and quantitative analysis, which provided retention time and characteristic fragments ratio for each pesticide so as to positively identify each analyte. Relative standard deviations (RSDs) were within 7.7% (5.0 - 22.5 µg/kg, n = 3). The recoveries of pesticide standards at the spiked concentration of 5.0 - 22.5 µg/kg were between 87.1% and 110.9%. Limits of detection (LODs) for the analytes were 0.16 - 3.2 µg/kg, which could meet the demand of routine analysis and TCM quality control.

INTRODUCTION

Traditional Chinese Medicine (TCM) usually means medicinal plants. Herb water, mixed tablet, and powder from the extracts of herbs are the general styles in clinical practice. The usage of TCM in China has a very long history. The influence of TCM on health care system has been profound in Asia as well as in the West in recent years [1-3]. The concerning of contaminations of heavy metals [4-6] and pesticides [2,6-9], which may be introduced during the cultivation, transportation, preparation and preservation, has been increased as the popularity of TCM enlarged.

Pesticides using to control various insect pests all over the world have advanced agriculture to gain great productivity [10, 11]. In the mean time, they contaminate the environment [12] and endanger human health [13]. Some pesticides are neural destroyers [14], and some act as hormones, which may disturb human endocritic system [15]. Most of the pesticides are bio-accumulated and may be transferred along the food chain [16], similar to the environmental behavior of heavy metals. For these reasons, the contents of pesticide residues in Chinese herbs have been concerning by public in China and the other areas of the world. There have methods reported for the analysis of pesticide residues in TCM [9, 17, 18]. However, a rapid procedure or screening method to determine organochlorine, organophosphorus, and pyrethroid pesticides in TCM, especially for complex matrix medicinal herbs, such as flos lonicerae, is of great significance in quality control activities.

Typically, measurement of multi-residue in complex matrices comprises sample pre-treatment, separation and detection by gas chromatography-mass spectrometry (GCMS) [19, 20]. The pre-treatment usually is time-consuming, which includes extraction, purification, and enhancement of the analytes to reduce or eliminate possible interference to the accurate detection. Soxhlet extraction [21], accelerated solvent extraction (ASE) [16], supercritical fluid extraction (SFE) [17], solid phase extraction (SPE) [12, 13], microwave-assisted extraction (MAE) [22], matrix solid phase disperse extraction (MSPDE) [19,23], and disperse solid phase extraction (DSPE, QuEChERS) [24-26] have been investigated for pesticides analysis. Gel permeation chromatography (GPC), as reported in bibliography [10,16,27,28], may be one of the best techniques for the analysis of multi-residue of pesticides in TCM, which separates lower molecular weight target pesticides from higher molecular weight chemical matrices, such as pigments. Gas chromatography-mass spectrometry can do two-tier identification and confirmation with the retention time and the relative ratios of characteristic ions of the pesticide. The select ion monitor (SIM) can eliminate matrix influence and enhance selectivity and sensitivity effectively. The double mode of scan plus SIM mode (scan/SIM) can qualify and quantify target compounds simultaneously in a single injection.

The purpose of this study was to develop a novel method for accurately and simultaneously determination of organochlorine, organophosphorus, and pyrethroid pesticides in flos lonicerae. The advantage of GPC purification and good separation and high sensitivity of GC-MS was investigated in qualitative identification and quantitative detection of multi-pesticide in complex chemical matrices.

EXPERIMENTAL

Chemicals and Reagents

Acetone, acetonitrile, cyclohexane and ethyl acetate were all of HPLC grade and purchased from Thermal Fisher Co. (USA). The 23-pesticide standards were from Chem Service Co. (USA), which were of purity ≥ 98.1%.

1000 µg/mL stock solutions for each pesticide were prepared with acetone and stored in freezer at −18°C. 10 µg/mL mixed standard solution for daily work were obtained by mixing and diluting stock solutions with acetone, and stored in a refrigerator at 4°C.

Apparatus

Agilent 7890A gas chromatography, 5975C mass spectrometer, and 7683B auto sample injector was used (Agilent Technologies, USA). Data acquisition, data processing, and instrumental control were performed with Agilent Enhanced ChemStation (Agilent Technologies, USA). A DB-5 MS fused silica capillary column of 30 m × 0.25 mm with 0.25 µm film thickness from Agilent Technologies was used.

The GPC system (Vario, LC Tech, Germany) consisted of high pressure pump, auto-sampler with 24-sample vials (10 mL) and 5.0 mL sample loop, GPC column, and 24 fraction collected vials (100 mL). The clean-up GPC column was packed polystyrene-divinylbenzene (Bio-Beads S-X3, 400 mm × 25 mm I.D., 200 - 400 mesh).

MS2 mini-shaker (IKA, Germany), B5200S-OT ultrasonic extract device (Branson, USA), centrifuge (Shanghai, China), auto concentrating apparatus (EVA Ш, LC Tech, Germany), and laboratory-built nitrogen evaporator were used.

Sample Extraction and Purification

1.000 g of ~2 kg homogenized dry flos lonicerae sample was extracted with 5.0 mL mixed ethyl acetate and cyclohexane (1:1, v/v) for 10 min. The extraction was repeated for 3 times. Combine the extracts together, and then concentrated with a nitrogen stream to 10.0 mL. 5.0 mL of the extract solution was injected and separated on GPC with a mixed mobile phase of ethyl acetate and cyclohexane (1:1, v/v) at a flow rate of 5.0 mL/min. The fraction containing the analyzed pesticides was collected within the retention time of 17 to 36 min (totally 95.0 mL of eluant). The GPC fraction was evaporated and concentrated to 5.0 mL using EVA Ⅲ rotary evaporator at 40°C, and further concentrated to near dry with nitrogen stream, re-dissolved with 0.5 mL mixture of ethyl acetate and cyclohexane (1:1, v/v) for GC-MS analysis.

GC-MS Conditions

Carrier gas was high purity of Helium (≥99.999%). The separation of all pesticides was performed with a constant pressure mode. The injection port of GC was 250°C. 1.0 μL of pretreated sample solution was injected in splitless mode (split valve closed for 0.75 min). The retention time was locked by chlorpyrifos-methyl. Programmed temperature for GC oven was initially 50°C for 1 min, increased to 125°C at a rate of 25°C/min, and then to 300°C at 10°C/min, and finally maintained at 300°C for 10 min until all the analytes eluted.

Electron impact (EI) ionization source was used at 70 eV. The interface between the GC and mass detector was maintained at 280°C. The temperature for EI source and quadrupole were set at 230°C and 150°C, respectively. The solvent delay time was set to 3.5 min. Full scan and selected ion mode (Scan/SIM) were used to qualitative identification and quantitative detection of multi-pesticide.

Figure 1 is the chromatograms of 23-pesticide standard and standard spiked flos lonicerae sample at optimized operating conditions. All pesticides in Figure 1 have been separated and eluted before the retention time 23.237 min. Additional 8 min at 300°C was set for column cleaning-up and ready for next injection

RESULTS AND DISCUSSION

Optimization of Extraction and Purification Procedure

The pesticides studied in this work are refereed to different natures, classes and physicochemical properties, especially their wide range of polarity. In fact, the sample pretreatment for multi-residue measurement is of difficulty. In this experiment, acetonitrile, cyclohexane, ethyl acetate, and mixture of ethyl acetate and cyclohexane (1:1, v/v) were investigated to be extract solvent to extract pesticides from flos lonicerae sample.

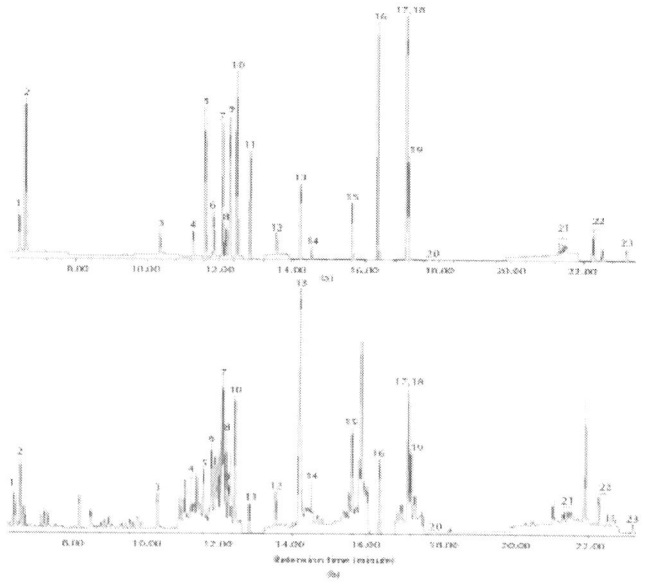

Figure 1: Chromatograms of 23-pesticide standard (a) and standard spiked flos lonicerae sample; (b) Peak identification: 1, Methamidophos; 2, Dichlorvos; 3, Omethoate; 4, Monocrotophos; 5, a-BHC; 6, Dimethoate; 7, b-BHC; 8, Quintozene; 9, Lindane; 10, Diazinon; 11, d-BHC; 12, Methyl parathion; 13, Malathion; 14, Parathion; 15, Methidathion; 16, p,p'-DDE; 17, Ethion; 18, p,p'-DDD; 19, o,p'-DDT; 20, p,p'-DDT; 21, Cypermethrin; 22, Fenvalerate; 23, Deltamethrin.

As one of the most efficient extract solvent, acetonitrile can yield more co-extracts and make the followed clean-up or purification step sophisticated. Furthermore, acetonitrile can not dissolve very well in cyclohexane which was used as the GPC mobile phase in this work. The extract dissolved within acetonitrile had to be evaporated and concentrated near to dry and then re-dissolved by GPC mobile phase before loading to purification when acetonitrile was used as extract solvent in this case. At the same time, non-polar cyclohexane was not good enough to extract most of the polar pesticides, such as organophosphorus pesticides. In the experiment, finally we found that ethyl acetate and ethyl acetate-cyclohexane mixture (1:1, v/v) were the most efficient extract solvent for all pesticides investigated. To match the GPC mobile phase, ethyl acetate-cyclohexane mixture (1:1, v/v) was used as the extract solvent in following experiments, which has proper polarity and can improve extraction efficiency and minimize matrix interferences.

The volume of GPC fraction, collected as the purified pesticides, was one of the most important operating parameters. In this work, the experiment results showed that most of the pesticides eluted in the retention time of 16 - 36 min. In the meantime, many pigments eluted in 16 - 17 min, which would had interference on the identification, certification, and determination of pesticides with GC-MS. Therefore, the GPC fraction in the retention time of 17 - 36 min, totally 95.0 mL, were collected and concentrated to 0.50 mL for GC-MS analysis.

GC-MS Detection

In this study, pesticide residue, usually at trace or ultratrace level, accompanied with complex matrices, although there had GPC clean-up step before analysis by GC-MS. The identification and certification of analyzed pesticides should be careful and take more evidences to avoid possible mistakes. The efficient separation of pesticides and continuum components by GC was essential in this work, which provided the retention times to identify analytes. In the same time, the capability of discerning characteristic fragments of each pesticide by MS detector could avoid or eliminate false positives in measurement. 4 characteristic fragment ions for each pesticide were chosen to calculate the ratio of characteristic ion abundance ratio. At last, the qualitative analysis, or the identification and certification, of all pesticides were performed with both the retention time and the ratios of the abundance of characteristic ions of each pesticide.

For the quantitative analysis of pesticides, the simultaneous full scan and selected ion monitor (scan/SIM) mode were used, which not only provide pesticide structure information but also improve the selectivity. Owe to the benefit of this scan/SIM mode, qualification and quantification can be completed synchronously in a single injection. Quantitative analysis of all pesticides was carried out with the abundance of a carefully selected characteristic fragment ion, which was free of the matrix interference. Table 1 listed all the 14 groups of monitoring ions for the 23 pesticides. Table 2 shown the retention time (t_r), quantifying ions, qualifying ions, and the abundance ratios of all characteristic ions for each pesticide. The uncertainty of ion ratios for qualification was controlled to lower than 20%.

Analytical Performance for the Developed Method

The developed method was validated with the recoveries of the spiked standards in flos lonicerae sample. In the experiments, we chose a pesticide free flos lonicerae sample as the chemical matrix. By spiking different concentration level of pesticides standard in sample matrices and aging in a refrigerator for at least 4 hr at 4°C, artificial samples were prepared to the evaluation of method accuracy, precision, calibration, limits of detection (LODs), and limits of quantification (LOQs) in following experiments.

Table 1: Monitoring ions segments of pesticides by GC-MS

Segments	Start time (minute)	Monitored ions (m/z)
1	3.397	94, 95, 141, 47, 109, 185, 79, 187
2	9.608	156, 110, 79, 109
3	10.882	127, 192, 67, 97, 181, 219, 183, 217, 87, 93, 125, 143, 219, 181, 183, 214, 181, 183, 219, 111, 179, 137, 152, 199
4	12.632	181. 219. 183.217

5	13.231	263. 109, 125, 79
6	13.884	173. 127, 125, 79
7	14.329	291. 109, 97 139
8	15.095	145. 85. 93. 125
9	15.999	246. 318. 316. 248
10	16.734	231, 153, 97,125, 235, 237, 165, 236, 235, 337, 165, 236
11	17.479	235, 237, 165. 236
12	19.836	181, 163, 165 77 181 163 165 209, 163, 181. 165. 209, 163, 181, 165, 209
13	21.859	167, 125. 181. 152. 167, 125, 181, 169
14	22.779	181, 253 251 255

Table 2: GC-MS parameters for determination 23-pesticide residues in flos lonicerae

No.	Compounds	Retention time (t, min)	Quantify ion (m/z)	Qualify ion	
				m/z	Relative abundance (%)
1	Methamidophos	6.430	94	94, 95, 141, 47	100, 62, 53, 17
2	Dichlorvos	6.595	109	109, 185, 79, 187	100, 33, 16, 11
3	Omethoate	10.348	156	156, 110, 79, 109	100, 84, 22, 21
4	Monocrotophos	11.281	127	127, 192, 67, 97	100, 16, 15, 15
5	BHC alpha isomer	11.613	181	181, 219, 183, 217	100, 96, 95, 75
6	Dimethoate	11.842	87	87, 93, 125, 143	100, 60, 59, 13
7	BHC beta isomer	12.097	219	219, 181, 183, 217	100, 99, 95, 79
8	Quintozene	12.181	237	237, 249, 295, 214	100, 76, 74, 66
9	Lindane	12.298	181	181, 183, 219, 111	100, 96, 87, 53
10	Diazinon	12.475	179	179, 137, 152, 199	100, 96, 65, 58
11	BHC delta isomer	12.853	181	181, 219, 183, 217	100, 98, 96, 78
12	Methyl parathion	13.570	263	263, 109, 125, 79	100, 95, 80, 23
13	Malathion	14.228	173	173, 127, 125, 93	100, 85, 83, 62

14	Parathion	14.531	291	291, 109, 97, 139	100, 78, 66, 44
15	Methidathion	15.627	145	145, 85, 93, 125	100, 59, 16, 15
16	p.p'-DDE	16.344	246	246, 318, 316, 248	100, 86, 67, 66
17	Ethion	17.125	231	231, 153, 97, 125	100, 51, 42, 35
18	p.p'-DDD	17.139	235	235, 237, 165, 236	100, 65, 41, 15
19	o.p'-DDT	17.191	235	235, 237, 165, 236	100, 66, 36, 15
20	p.p'-DDT	17.851	235	235, 237, 165, 236	100, 66, 35, 15
21	Cypermethrin I	21.359	181	181, 163, 165, 77	100, 89, 76, 34
	Cypermethrin II	21.461	181	181, 163, 165, 209	100, 94, 79, 38
	Cypermethrin III	21.516	163	163, 181, 165, 209	100, 81, 66, 45
	Cypermethrin IV	21.555	163	163, 181, 165, 209	100, 81, 66, 46
22	Fenvalerate I	22.317	167	167, 125, 181, 152	100, 97, 62, 55
	Fenvalerate II	22.553	167	167, 125, 181, 169	100, 99, 62, 54
23	Deltamethrin	23.237	181	181, 253, 251, 255	100, 67, 43, 33

Accuracy and Precision

Accuracy and precision were performed by the standard recovery test with artificial flos lonicerae sample. The standard recovery test was carried out at three levels of spiking concentration for each pesticide. Each spiking level was repeated three times to evaluate the precision. Accuracy was assessed by the standard added recoveries, and precision by the relative standard deviations (RSDs). Table 3 summarizes the standard added recoveries and RSDs for developed method.

The result in Table 3 showed that standard added recoveries were in the range of 87.1% - 110.9% at low spiked level (5.0 - 22.5 µg/kg), 86.4% - 107.1% at medium spiked level (8.2 - 45.0 µg/kg), and 71.9% - 108.5% at high spiked level (41.2 - 224.9 µg/kg). The linear regressions of the standard added concentration (x) versus peak area (y) of quantifying ion for all analytes yielded the correlation coefficients closing to 1.000, as shown in Table 3. These results suggest that the standard added recoveries for all pesticides are stable at low, medium, and high concentration, which made the developed method more practical and useful in real world sample analysis.

RSDs obtained were lower than 10% at all three added concentration levels, with the exception of 22.0% for p, p›-DDT at a high level of 214.9 µg/kg which may be interfered by the sample matrices. These results show that the developed method has a good reproducibility in the added concentration range.

Calibration

Matrix enhancement or reduction effect on the signal response is one of the most common problems in trace pesticides analysis for complex matrix samples [16, 29]. In this work, the quantifications of all pesticides were carried out with matrix-matched external standard calibration method.

Table 3: Recovery, relative standard deviation and correlation coefficient of 23 pesticides in flos lonicerae by GPC-GC-MS

No	Compounds	Low spiked level			Medium spiked level			High spiked level			Correlation coefficient
		Concentration (µg/kg)	Recovery (%)	RSD (%)	Concentration (µg/kg)	Recovery (%)	RSD (%)	Concentration (µg/kg)	Recovery (%)	RSD (%)	
1	Methamidophos	22.0	93.9	2.8	44.0	86.4	9.1	219.7	85.1	2.4	0.9999
2	Dichlorvos	20.3	81.0	3.7	40.6	87.7	6.8	203.0	71.9	5.8	0.9982
3	Omethoate	22.4	89.2	3.1	44.9	94.6	1.6	224.4	101.5	4.7	0.9998
4	Monocrotophos	20.5	90.6	2.7	41.0	96.7	6.9	205.0	104.5	3.2	0.9991
5	BHC alpha isomer	19.9	102.1	6.2	39.7	96.6	7.2	198.6	99.2	4.7	0.9994
6	Dimethoate	20.8	98.8	6.6	41.6	103.9	5.0	207.9	105.4	3.1	0.9998
7	BHC beta isomer	20.8	102.5	1.1	41.6	101.6	6.7	207.9	104.4	2.5	0.9994
8	Quintozene	21.3	98.7	6.0	42.7	96.9	7.4	213.4	98.0	5.3	0.9998
9	Lindane	22.5	102.2	4.0	45.0	97.4	6.3	224.9	101.8	3.6	0.9997
10	Diazinon	22.2	110.9	5.1	44.4	103.2	9.6	222.2	102.0	2.2	0.9990
11	BHC delta isomer	20.1	102.1	6.8	40.2	100.2	9.1	200.8	103.5	2.7	0.9996
12	Methyl parathion	22.5	102.0	6.8	45.0	105.6	9.2	224.9	106.4	3.0	1.0000
13	Malathion	22.1	102.6	4.8	44.2	107.1	2.1	220.9	101.9	1.2	0.9966
14	Parathion	22.5	103.2	3.9	45.0	102.9	7.0	224.9	102.3	3.0	0.9999
15	Methidathion	19.9	101.0	3.9	39.7	99.1	4.6	198.6	99.7	2.8	0.9997
16	p,p'-DDE	20.0	108.9	2.3	40.0	96.6	8.3	200.2	99.7	3.0	0.9997
17	Ethion	22.1	105.1	3.7	44.2	99.8	7.6	221.3	97.5	2.1	0.9999
18	p,p'-DDD	20.2	107.8	1.3	40.4	101.3	4.6	201.9	98.9	2.5	0.9996
19	p,p'-DDT	21.8	107.0	2.6	45.7	100.2	6.9	218.2	99.3	3.0	0.9998
20	p,p'-DDT	21.5	87.1	7.7	45.0	101.7	4.6	214.9	108.5	22.0	0.9712
21	Cypermethrin I	5.4	100.9	3.9	10.7	97.0	2.9	53.6	98.5	3.5	0.9999
	Cypermethrin II	5.0	96.6	2.4	10.0	88.3	4.8	49.7	94.3	3.3	1.0000
	Cypermethrin III	_a	-	-	13.1	93.2	7.9	65.3	98.1	1.7	-
	Cypermethrin IV	-	-	-	8.2	91.0	5.2	41.2	97.3	5.6	-
22	Fenvalerate I	15.2	91.8	2.9	30.5	97.9	9.1	152.3	98.7	3.6	0.9999
	Fenvalerate II	7.5	97.3	2.4	14.9	88.5	4.6	74.6	98.4	3.9	1.0000
23	Deltamethrin	20.3	95.0	3.8	41.0	89.4	9.5	203.8	98.5	3.7	1.0000

ᵃNo data available.

The calibration curves for quantification were obtained by measurement of a series of mixed pesticide standards. The standard series had 7 different concentration level between 10.0 and 1000.0 $\mu g \cdot L^{-1}$ as listed in Table 4. The linear curves were plotted by least squares regression of concentration versus peak area of each pesticide. The linear ranges and correlation coefficients were shown in Table 4.

Limits of Detection (LODs) and Limits of Quantification (LOQs)

The measurement of an artificial flos lonicerae sample, mixed standard added at 20 $\mu g \cdot kg^{-1}$, was repeated three times, including the pretreatment, extract, GPC purification, and GC-MS detection. The standard derivations (SDs) of the measurement for each pesticide were then calculated. Here Limits of detections (LODs) of the developed method were defined as the concentrations corresponding to 3 times of the standard deviations (SDs) of each analyte. LOD for each pesticide was in the range of 0.16 - 3.24 $\mu g \cdot kg^{-1}$, with the exception of malathion and p,p'- DDT which had no data available.

CONCLUSIONS

A method for simultaneous measurement of 23 organochlorine, organophosphorus, and pyrethroid pesticides in traditional Chinese medicine, flos lonicerae, using GPC purification and GC-MS detection was developed. The method has good accuracy and precision, as well as low LODs. The method showed great prospects in determining common classes of pesticides in a typical traditional Chinese medicine.

Table 4: Linear ranges, correlation coefficients, LODs and LOQs for 23 pesticides in flos lonicerae

No.	Compounds	Linear range (μg/L)	Correlation coefficient	LOD (μg/kg)	LOQ (μg/kg)
1	Methamidophos	22 - 1099	0.9992	0.96	3.18
2	Dichlorvos	10 - 1015	0.9969	1.05	3.49
3	Omethoate	11 - 1122	0.9984	0.82	2.72
4	Monocrotophos	10 - 1025	0.9983	1.54	5.13
5	BHC alpha isomer	10 - 993	0.9976	1.95	6.52
6	Dimethoate	10 - 1040	0.9994	2.02	6.74
7	BHC beta isomer	10 - 1040	0.9985	0.40	1.35
8	Quintozene	11 - 1067	0.9995	2.70	9.00
9	Lindane	11 - 1124	0.9981	1.20	3.99
10	Diazinon	11 - 1111	0.9986	2.18	7.28
11	BHC delta isomer	10 - 1004	0.9985	2.35	7.83
12	Methyl parathion	11 - 1124	0.9972	3.24	10.81
13	Malathion	11 - 1104	0.9983	[a]	-
14	Parathion	11 - 1124	0.9939	1.45	4.83
15	Methidathion	10 - 993	0.9989	1.09	3.65
16	p,p'-DDE	10 - 1000	0.9987	0.83	2.77
17	Ethion	11 - 1105	0.9992	1.30	4.34
18	p,p'-DDD	10 - 1010	0.9993	0.43	1.44
19	p,p'-DDT	11 - 1091	0.9990	0.92	3.07
20	p,p'-DDT	54 - 1075	0.9912	-	-
21	Cypermethrin I	3 - 268	0.9988	0.35	1.15
	Cypermethrin II	3 - 249	0.9988	0.16	0.53
	Cypermethrin III	16 - 327	0.9966	-	-
	Cypermethrin IV	10 - 206	0.9938	-	-
22	Fenvalerate I	8 - 761	0.9996	0.61	2.03
	Fenvalerate II	4 - 373	0.9997	0.20	0.67
23	Deltamethrin	10 - 1024	0.9996	2.09	6.98

[a]No data available.

ACKNOWLEDGEMENTS

This study is funded by the Natural Science Foundation of Fujian Province, PR China (B0220001). The authors thank Dr. Dengyun CHEN and all colleagues in the research group of Marine Analytical Chemistry for their assistance. Thanks also to the editor and anonymous for their comments that greatly improved the quality of the paper.

REFERENCES

1. K. Chan, "Progress in Traditional Chinese Medicine," Trends in Pharmacological Sciences, Vol. 16, No. 6, 1995, pp. 182-187. doi:10.1016/S0165-6147(00)89019-7

2. K. Chan, "Some Aspects of Toxic Contaminants in Herbal Medicines," Chemosphere, Vol. 52, No. 9, 2003, pp. 1361-1371. doi:10.1016/S0045-6535(03)00471-5

3. P. Drasar and J. Moravcova, "Recent Advances in Analysis of Chinese Medical Plants and Traditional Medicines," Journal of Chromatography B: Analytical Technologies in the Biomedical and Life Sciences, Vol. 812, No. 1-2, 2004, pp. 3-21. doi:10.1016/S1570-0232(04)00741-X

4. E. Ernst, "Toxic Heavy Metals and Undeclared Drugs in Asian Herbal Medicines," Trends in Pharmacological Sciences, Vol. 23, No. 3, 2002, pp. 136-139. doi:10.1016/S0165-6147(00)01972-6

5. R. J. Huang, Z. X. Zhuang, Y. Tai, R. F. Huang, X. R. Wang and F. S. C. Lee, "Direct Analysis of Mercury in Traditional Chinese Medicines Using Thermolysis Coupled with Online Atomic Absorption Spectrometry," Talanta, Vol. 68, No. 3, 2006, pp. 728-734.doi:10.1016/j.talanta.2005.05.014

6. F. M. Li, Z. L. Xiong, X. M. Lu, F. Qin and X. Q. Li, "Strategy and Chromatographic Technology of Quality Control for Traditional Chinese Medicines," Chinese Journal of Chromatography, Vol. 24, No. 6, 2006, pp. 537- 544. doi: 10.1016/S1872-2059(06)60022-9

7. J. F. Deng, "Clinical and Laboratory Investigations in Herbal Poisonings," Toxicology, Vol. 181, 2002, pp. 571- 576. doi:10.1016/S0300-483X(02)00484-5

8. J. Jung, M. Hermanns-Clausen and W. Weinmann, "Anorectic Sibutramine Detected in a Chinese Herbal Drug for Weight Loss," Forensic Science International, Vol. 161, No. 2-3, 2006, pp. 221-222. doi:10.1016/j.forsciint.2006.02.052

9. J. Xue, L. L. Hao and F. Peng, "Residues of 18 Organochlorine Pesticides in 30 Traditional Chinese Medicines," Chemosphere, Vol. 71, No. 6, 2008, pp. 1051- 1055.doi:10.1016/j.chemosphere.2007.11.014

10. A. G. Frenich, P. P. Bolanos and J. L. M. Vidal, "Multiresidue Analysis of Pesticides in Animal Liver by Gas Chromatography Using Triple Quadrupole Tandem Mass Spectrometry," Journal of Chromatography A, Vol. 1153, No. 1-2, 2007, pp. 194-202. doi:10.1016/j.chroma.2007.01.066

11. S. N. Sinha and M. Odetokun, "Liquid Chromatography Mass Spectrometer (LC-MS/MS) Study of Distribution Patterns of Base Peak Ions and Reaction Mechanism with Quantification of Pesticides in Drinking Water Using a Lyophilization Technique," American Journal of Analytical Chemistry, Vol. 2, No. 5, 2011, pp. 511-521.doi:10.4236/ajac.2011.25061

12. S. Wang, P. Zhao, G. Min and G. Z. Fang, "Multi-Residue Determination of Pesticides in Water Using MultiWalled Carbon Nanotubes SolidPhase Extraction and Gas Chromatography-Mass Spectrometry," Journal of Chromatography A, Vol. 1165, No. 1-2, 2007, pp. 166-171. doi:10.1016/j.chroma.2007.07.061

13. P. Liu, Q. Liu, Y. Ma, J. Liu and X. Jia, "Analysis of Pesticide Multiresidues in Rice by Gas ChromatographyMass Spectrometry Coupled with Solid Phase Extraction," Chinese Journal of Chromatography, Vol. 24, No. 3, 2006, pp. 228-234. doi:10.1016/ S1872-2059(06)60011-4

14. C. S. Roegge, O. A. Timofeeva, F. J. Seidler, T. A. Slotkin and E. D. Levin, "Developmental Diazinon Neurotoxicity in Rats: Later Effects on Emotional Response," Brain Research Bulletin, Vol. 75, No. 1, 2008, pp. 166- 172.doi:10.1016/j. brainresbull.2007.08.008

15. R. McKinlay, J. A. Plant, J. N. B. Bell and N. Voulvoulis, "Calculating Human Exposure to Endocrine Disrupting Pesticides Via Agricultural and Non-Agricultural Exposure Routes," Science of the Total Environment, Vol. 398, No. 1-3, 2008, pp. 1-12. doi:10.1016/j.scitotenv.2008.02.056

16. A. G. Frenich, J. L. M. Vidal, A. D. C. Sicilia, M. J. G. Rodriguez, P. P. Bolanos and R. G. A. C. Contaminan, "Multiresidue Analysis of Organochlorine and Organophosphorus Pesticides in Muscle of Chicken, Pork and Lamb by Gas Chromatography-Triple Quadrupole Mass Spectrometry," Analytica Chimica Acta, Vol. 558, No. 1-2, 2006, pp. 42-52. doi:10.1016/j.aca.2005.11.012

17. Y. C. Ling, H. C. Teng and C. Cartwright, "Supercritical Fluid Extraction and Clean-up of Organochlorine Pesticides in Chinese Herbal Medicine," Journal of Chromatography A, Vol. 835, No. 1-2, 1999, pp. 145-157. doi:10.1016/S0021-9673(98)01077-2

18. M. Barriada-Pereira, E. Concha-Grana, M. J. GonzalezCastro, S. Muniategui-Lorenzo, P. Lopez-Mahia, D. PradaRodriguez and E. Fernandez-Fernandez, "MicrowaveAssisted Extraction Versus Soxhlet Extraction in the Analysis of 21 Organochlorine Pesticides in Plants," Journal of Chromatography A, Vol. 1008, No. 1, 2003, pp. 115- 122.doi:10.1016/S0021-9673(03)01061-6

19. P. P. Bolanos, A. G. Frenich and J. L. M. Vidal, "Application of Gas Chromatography-Triple Quadrupole Mass Spectrometry in the Quantification-Confirmation of Pesticides and Polychlorinated Biphenyls in Eggs at Trace Levels," Journal of Chromatography A, Vol. 1167, No. 1, 2007, pp. 9-17. doi:10.1016/j. chroma.2007.08.019

20. Z. Q. Huang, Y. J. Li, B. Chen and S. Z. Yao, "Simultaneous Determination of 102 Pesticide Residues in Chinese Teas by Gas Chromatography-Mass Spectrometry," Journal of Chromatography B-Analytical Technologies in the Biomedical and Life Sciences, Vol. 853, No. 1-2, 2007, pp. 154-162. doi:10.1016/j. jchromb.2007.03.013

21. S. H. Hong, U. H. Yim, W. J. Shim, J. R. Oh, P. H. Viet and P. S. Park, "Persistent Organochlorine Residues in Estuarine and Marine Sediments from Ha Long Bay, Hai Phong Bay, and Ba Lat Estuary, Vietnam," Chemosphere, Vol. 72, No. 8, 2008, pp. 1193-1202. doi:10.1016/j.chemosphere.2008.02.051

22. S. B. Singh, G. D. Foster and S. U. Khan, "Determination of Thiophanate Methyl and Carbendazim Residues in Vegetable Samples Using Microwave-Assisted Extraction," Journal of Chromatography A, Vol. 1148, No. 2, 2007, pp. 152-157. doi:10.1016/j.chroma.2007.03.019

23. M. G. D. Silva, A. Aquino, H. S. Drea and S. Navickiene, "Simultaneous Determination of Eight Pesticide Residues in Coconut Using Mspd and Gc/Ms," Talanta, Vol. 76, No. 3, 2008, pp. 680-684. doi:10.1016/j.talanta.2008.04.018

24. C. Lesueur, P. Knittl, M. Gartner, A. Mentler and M. Fuerhacker, "Analysis of 140 Pesticides from Conventional Farming Foodstuff

Samples after Extraction with the Modified Quechers Method," Food Control, Vol. 19, No. 9, 2008, pp. 906-914.doi:10.1016/j. foodcont.2007.09.002

25. T. D. Nguyen, E. M. Han, M. S. Seo, S. R. Kim, M. Y. Yun, D. M. Lee and G. H. Lee, "A Multi-Residue Method for the Determination of 203 Pesticides in Rice Paddies Using Gas Chromatography/ Mass Spectrometry," Analytica Chimica Acta, Vol. 619, No. 1, 2008, pp. 67-74. doi:10.1016/j.aca.2008.03.031

26. T. D. Nguyen, J. E. Yu, D. M. Lee and G. H. Lee, "A Multiresidue Method for the Determination of 107 Pesticides in Cabbage and Radish Using Quechers Sample Preparation Method and Gas Chromatography Mass Spectrometry," Food Chemistry, Vol. 110, No. 1, 2008, pp. 207-213. doi:10.1016/j.foodchem.2008.01.036

27. B. Hu, W. Song, L. Xie and T. Shao, "Determination of 33 Pesticides in Tea Using Accelerated Solvent Extraction/Gel Permeation Chromatography and Solid Phase Extraction/ Gas Chromatography-Mass Spectrometry," Chinese Journal of Chromatography, Vol. 26, No. 1, 2008, pp. 22-28. doi:10.1016/ S1872-2059(08)60009-7

28. F. Goni, R. Lopez, A. Etxeandia, E. Millan, A. Vives and P. Amiano, "Method for the Determination of Selected Organochlorine Pesticides and Polychlorinated Biphenyls in Human Serum Based on a Gel Permeation Chromatographic Clean-Up," Chemosphere, Vol. 76, No. 11, 2009, pp. 1533-1539. doi:10.1016/j. chemosphere.2009.05.041

29. C. Ferrer, M. J. Gomez, J. F. Garcia-Reyes, I. Ferrer, E. M. Thurman and A. R. Fernandez-Alba, "Determination of Pesticide Residues in Olives and Olive Oil by Matrix Solid-Phase Dispersion Followed by Gas Chromatography/Mass Spectrometry and Liquid Chromatography/ Tandem Mass Spectrometry," Journal of Chromatography, A, Vol. 1069, No. 2, 2005, pp. 183-194. doi:10.1016/j.chroma.2005.02.015

Chapter

3

Vehicular Quality Biomethane Production from Biogas by Using an Automated Water Scrubbing System

R. Chandra[1], V. K. Vijay[2], and P. M. V. Subbarao[3]

[1]Department of Farm Power & Machinery, College of Agricultural Engineering & Post Harvest Technology, Central Agricultural University, Ranipool, Gangtok, Sikkim 737 135, India

[2]Centre for Rural Development & Technology, Indian Institute of Technology Delhi, Hauz Khas, New Delhi 110 016, India

[3]Department of Mechanical Engineering, Indian Institute of Technology Delhi, Hauz Khas, New Delhi 110 016, India

ABSTRACT

This paper presents the results of an automated water scrubbing system used for enrichment of methane content in the biogas, to produce vehicular grade biomethane fuel. Incorporation of automatic control systems for precisely regulating the water level and maintaining constant operating pressure in the packed bed absorption column of water scrubbing system resulted in steady-state operation of the scrubbing system and a consistent supply of methane-enriched biogas from the gas outlet. The improved automated water scrubbing system was found to enrich 97% methane at an operating column pressure of 1.0 MPa with 2.5 m^3/h biogas in-flow rate and 2.0 m^3/h water in-flow rate into the scrubbing column unit.

INTRODUCTION

The future energy security concerns along with the increasing concentration of the carbon dioxide and methane greenhouse gases emission problems have strengthened the interest in development and utilization of alternative, non-petroleum-based renewable sources of energy [1, 2]. Biogas is an important renewable fuel among the various biomass-derived renewable fuel, particularly for rural areas. It is an environment friendly, clean, cheap, and versatile fuel. In fact, all over the world, biogas has been extensively used for heating purposes and/or electricity generation [3].

The presence of high concentration of carbon dioxide in biogas lowers the energy content per unit mass/volume and limits its utility to only low quality energy applications. The presence of carbon dioxide in biogas is undesirable to use as a vehicular fuel because it lowers the power output from the engine and it occupies additional space in the storage cylinders. Hence, it reduces the refilling range of the vehicle. Furthermore, the presence of carbon dioxide in biogas can cause problem of freezing at valves and metering points. Therefore, removal of carbon dioxide from the biogas may enhance the utility of biogas for wider range of applications. With available technologies, it is economically possible to enrich methane content of biogas up to the natural gas level. The methane enrichment of biogas to biomethane quality and feeding into the natural gas grid/compression in cylinders is an effective way of integrating the biogas into the energy sector. Thus, it can be successfully used as substitute of natural gas and transportation fuel, combined heat and power, and electricity generation applications [4, 5].

In this present study/experiment, a water scrubbing system for methane enrichment in biogas was modified and automated using electronic control system. The electronic control system runs the system under steady state and maintains a consistent methane quality in purified gas outlet stream, in order to produce vehicular quality biomethane from biogas.

PROCESSES OF METHANE ENRICHMENT IN BIOGAS

Methane enrichment in biogas can be carried out by various processes. Table 1 presents the different methods/processes with their comparatives in term of various operational parameters.

Table 1: Comparison of different methods of methane enrichment in biogas

Sl. No.	Method	Advantages	Disadvantages
(1)	Absorption in water	One of the easiest and cheapest methods for CO_2 removal. Recommended for rural application.	Water pumping load is high and some loss of methane with washing water.
(2)	Absorption by chemicals	The chemical absorbents are more efficient in low pressure and can remove CO_2 to low partial pressures in treated gas.	Regeneration of the solvent requires a relatively high energy input. Disposal of by-product formed due to chemical reactions is a problem.

(3)	Pressure swing adsorption	By proper choice of the adsorbent, this process can remove CO_2, H_2S, moisture and other impurities.	Adsorption is accomplished at high temperature and pressure. Regeneration is carried out by vacuum. It is a costly process.
(4)	Membrane separation	Modular in nature and separate CO_2 and CH_4 effectively.	High pressure requirement. The processing cost is also high.
(5)	Cryogenic separation	Allows recovery of pure component in the form of liquid, which can be transported conveniently	High cost involved makes it impractical for biogas applications.
(6)	Chemical conversion	Extremely high purity in the product gas.	Process is extremely expensive and is not warranted in most cases of biogas applications.

Water Scrubbing of Biogas

Carbon dioxide has higher solubility in water than methane. Therefore, more carbon dioxide is dissolved in water than methane, particularly at lower temperature. In the water scrubbing process, carbon dioxide dissolves in the water, while the methane concentration in the gas phase increases. The gas leaving the scrubber has, therefore, an increased concentration of methane. Apart from carbon dioxide, water is capable of removing other impurities such as hydrogen sulfide, ammonia, hydrogen phosphide, chlorinated hydrocarbons, and others. The water leaving the absorption column contains dissolved gases mainly carbon dioxide with very little amount of methane and others. The operation of water scrubbing system is highly dependent on the solubility of carbon dioxide at particular temperature and pressure in water to form dilute carbonic acid

$$CO_2+H_2O=H_2CO_3 \qquad (1)$$

The compressed raw biogas is fed to a packed bed absorption column from the bottom of the scrubber, and pressurized water is sprayed from the top (counter-current flow of biogas and water). This process is one of simplest and cheapest method of methane enrichment in biogas/landfill gas for small as well as larger scales. Another advantage of water scrubbing system over other processes is that the out coming water containing dissolved carbon dioxide and other impurities from the scrubbing column is fairly easy to dispose. However, in other processes the chemicals used require special handling and disposal in order of environmental safety and health hazards [6–9].

MATERIALS AND METHODS

The water-scrubbing-based methane enrichment in biogas and further compression/bottling into CNG cylinder system was designed and developed at Indian Institute of Technology Delhi, India, for enhancing the utility of biogas application that is, vehicular use. The system consists a water scrubbing column and methane-enriched biogas compression system. Figure 1 shows the schematic diagram of complete system. The lack of automation in the developed system had resulted into inconsistent methane quality in purified gas outlet stream due to continuous fluctuation of water level, inlet biogas flow rate, and thereby operating pressure of scrubbing column. Therefore, in order to get a consistent methane-quality-enriched biogas, the existing water scrubbing system was automated by using electronic control/instrumentations to run the system under steady state. Figure 2 shows the automated water scrubbing system. Brief descriptions of the incorporated controls were as follows.

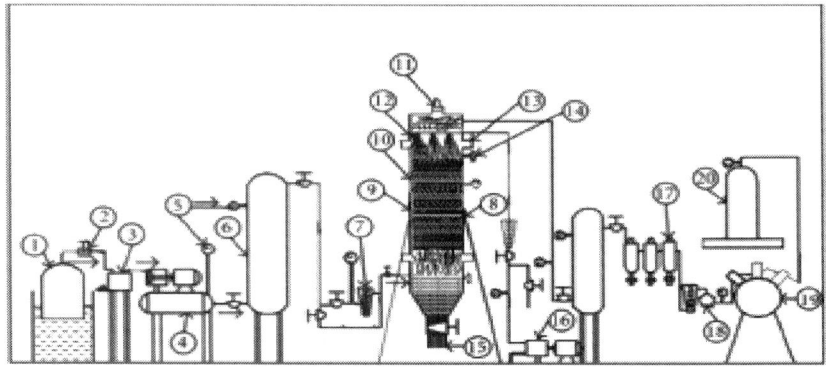

No.	Description	No.	Description	No.	Description
(1)	Biogas plant	(2)	Ball valve	(3)	Water remover
(4)	Compressor mounted with biogas receiver	(5)	Pressure gauge	(6)	Gas storage vessel
(7)	Rotameter	(8)	Supporting stand	(9)	Reshching rings
(10)	Scrubber	(11)	Pressure safety valve	(12)	Water sprayer
(13)	Flange	(14)	Water level view glass	(15)	CO_2 laden water outlet
(16)	High pressure water pump	(17)	Moisture removing filter	(18)	Pressure reducer
(19)	Three stage gas compressor	(20)	CNG cylinder		

Figure 1: Schematic diagram of water-scrubbing-based methane enrichment in biogas and compression system.

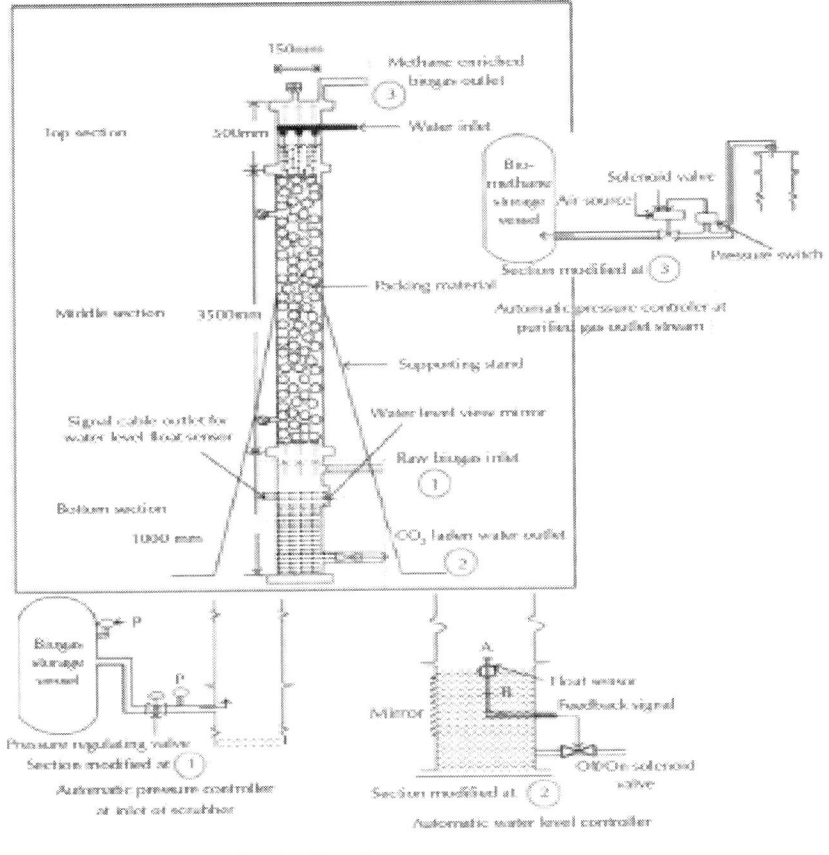

Figure 2: Incorporated automatic control systems in packed bed absorption column of water scrubbing system.

Automatic Pressure Controller at Inlet of Scrubber

A pressure regulating valve was fitted between the raw biogas storage vessel and input biogas line to the water-scrubbing column unit. The pressure regulating valve maintained almost constant pressure at the inlet of the water-scrubbing column unit irrespective of high pressure inside raw biogas storage pressure vessel.

Automatic Water Level Controller

An automatic water level controller was installed at the dissolved carbon dioxide laden water outlet of scrubbing column unit. This maintained a constant water level. The water level was maintained between two points A and B with the help of a "magnetic reed switch type float sensor" and "On/Off type solenoid valve." The float ball of the sensor moved according to the water level. When water level reached above the point "A" sensor generates a "normally open" signal to open the solenoid valve and water flows out till the water level set value and when the float ball reaches at point "B", then it generates a "normally closed" signal to off the solenoid valve.

Automatic Pressure Controller at Purified Gas Outlet Stream

The automatic pressure controller setup installed at the outlet of purified biogas line was able to maintain a constant pressure range of ~0.1 to 1.0 MP$_a$ at the outlet gas stream of the water scrubbing column unit. A differential pressure switch was configured to set the pressure value of 1.0 MP$_a$ and differential pressure value of 0.1 MP$_a$. Therefore, when the pressure in the scrubbing column reached a value of 1.0 MP$_a$, "pressure switch" generates open signal to the "solenoid valve," which further allowed the pneumatic source to open the "pneumatic ball valve" and released the gas to purified biogas storage vessel. As soon as the pressure in the scrubbing column unit decreased to a set value, "pressure switch" generates closes signal which closes the "solenoid valve" which further shuts off the "pneumatic ball valve" to stop the release of gas to purified biogas storage vessel. Thus, it had maintained constant pressure in the range of 0.9–1.0 MP$_a$ in water scrubbing column unit.

Working of the System

The raw biogas from a 20 m^3/day capacity floating drum type biogas plant fed with jatropha de-oiled seed cake was passed through a single-stage compressor after removing moisture and stored in a pressure vessel at more than 1.0 MP$_a$. The compressed raw biogas of desired flow rate was fed to the scrubbing column unit through a rotameter. A nonreturn valve was provided in the line to prevent back flow of the fed raw biogas. Biogas and water flow rates were regulated through valves, and a counter flow of gas and water was maintained in a scrubbing-column unit. At bottom section of the scrubber column, water level was maintained up to half mark with help of an automatic water level controller. It acts as a water seal and prevents escape of compressed biogas from the bottom of the scrubbing column. Carbon dioxide absorbed water was discharged through outlet of the scrubbing column. The gas coming out (methane enriched biogas) from the top of the scrubbing column was stored in a pressure vessel for further compression and storage in CNG cylinders and analyzed for methane and carbon dioxide contents.

System Performance Parameters

The performance of improved water scrubbing system was evaluated in terms of percentage of carbon dioxide absorbed in water (i.e., performance index). Methane and carbon dioxide contents in raw biogas were 65% and 32%, respectively. The effect of column-operating pressure, in-flow rates of water and biogas on quality of methane enriched biogas was studied. The operating variables were column-operating pressure, biogas in-flow, rate and water in-flow rate and are shown in Table 2. The performance index was computed by using

$$\sigma = (1-(y_e / y_r)) / (1-(y_e / 100)) \times 100 \tag{2}$$

Where σ is the performance index of scrubbing column, %; y_e is the volumetric content of carbon dioxide in methane enriched biogas, %; y_r is the volumetric content of carbon dioxide in raw biogas, %.

Table 2: Performance of automated water scrubbing system

Column pressure, MPa	Biogas in-flow rate, m3/h	Water in-flow rate, m3/h	CO2 remained in enriched biogas, %	Performance index ( ), %

0.8	1.0	1.5	5.0	88.8
	1.5	1.5	3.5	92.3
	2.0	1.5	4.0	91.1
	2.5	1.5	6.5	85.2
	3.0	2.0	8.5	80.3
1.0	1.0	1.5	4.0	91.1
	1.5	1.5	5.0	88.8
	2.0	1.5	3.5	92.3
	2.5	2.0	3.0	93.4
	3.0	2.5	5.0	88.8

RESULTS AND DISCUSSION

Table 2 shows the observed values of carbon dioxide content in methane-enriched biogas obtained from the purified gas-stream of the scrubbing unit for 0.8 MP$_a$ as well as 1.0 MP$_a$ operating column pressure.

Figures 3 and 4 depict the variation of the performance index with respect to biogas in-flow rate at column-operating pressure of 0.8 and 1.0 MP$_a$, respectively. It is evident from Table 2 and Figure 3 that the value of performance index first increased then decreased with increase in biogas in-flow rate at 0.8 MP$_a$ column-operating pressure. The highest performance index of 92.3% was observed at 1.5 m^3/h in-flow rates of biogas as well as water. Similarly, the highest value of performance index was observed as 93.4% at biogas in-flow rate of 2.5 m^3/h and water in-flow rate of 2.0 m^3/h with the column-operating pressure of 1.0 MP$_a$.

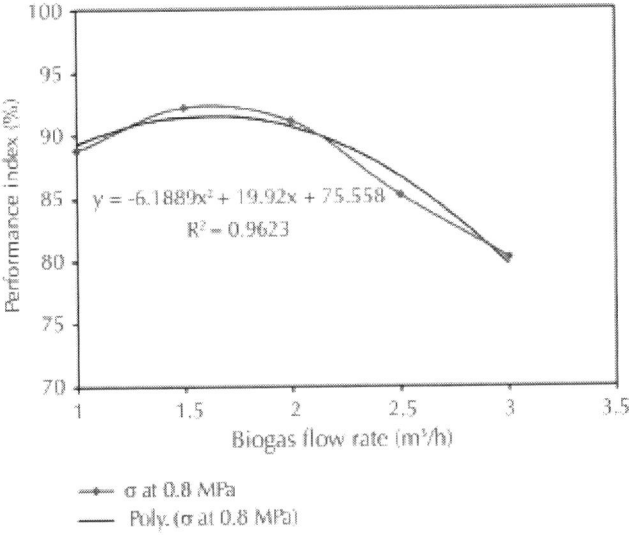

Figure 3: Effect of in-flow biogas rate on performance index at 0.8 MPa column-operating pressure.

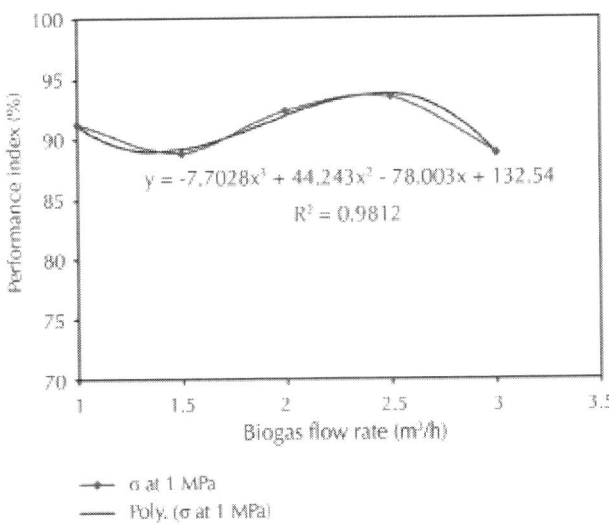

Figure 4: Effect of in-flow biogas rate on performance index at 1.0 MPa column-operating pressure.

The observed results showed that the percentage absorption of carbon dioxide increases with increase in scrubbing column pressure for all biogas in-flow rates. The highest carbon dioxide absorption of 90.6% was observed at 1.0 MP$_a$ column-operating pressure with a methane purity of 97% in purified gas outlet stream of the water-scrubbing system. The performance result of a 5.9 kW stationary diesel engine converted into spark ignition mode, run on compressed natural gas (CNG) and methane-enriched biogas (Bio-CNG) obtained from improved water scrubbing system, has showed that the engine performance is almost similar to that of compressed natural gas without any significant power loss [10].

CONCLUSIONS

The incorporation of automatic control systems in the water scrubbing unit has resulted into a steady state operation of the system which had provided a consistent methane quality in the outlet stream of purified gas. The improved automated water scrubbing unit enriched the biogas up to 97% methane at 1.0 MP$_a$ column-operating pressure with 2.5 m^3/h biogas in-flow rate and 2.0 m^3/h water in-flow rate. The incorporated automation in the system was found extremely satisfactory. Furthermore, the installation of automatic controls in the system had resulted into a significant labour savings for efficient operation of the biogas enrichment system. Further, the methane-enriched biogas is a good gaseous fuel as good as natural gas, and also biogas is renewable and CO$_2$ neutral fuel in terms of net emissions of carbon to the atmosphere.

ACKNOWLEDGMENTS

The authors are highly thankful to Centre for Rural Development and Technology and Mechanical Engineering Department, Indian Institute of Technology Delhi, New Delhi, India, for providing necessary facilities, support, and financial funding to conduct this paper.

REFERENCES

1. C. N. Hamelinck, G. Van Hooijdonk, and A. P. C. Faaij, "Ethanol from lignocellulosic biomass: techno-economic performance in short-, middle- and long-term," Biomass and Bioenergy, vol. 28, no. 4, pp. 384–410, 2005.

2. Y. Sun and J. Cheng, "Hydrolysis of lignocellulosic materials for ethanol production: a review,"Bioresource Technology, vol. 83, no. 1, pp. 1–11, 2002.

3. A. Hilkiah Igoni, M. J. Ayotamuno, C. L. Eze, S. O. T. Ogaji, and S. D. Probert, "Designs of anaerobic digesters for producing biogas from municipal solid-waste," Applied Energy, vol. 85, no. 6, pp. 430–438, 2008.

4. A. Demirbas, "Political, economic and environmental impacts of biofuels: a review," Applied Energy, vol. 86, no. 1, pp. S108–S117, 2009.

5. S. S. Kapdi, V. K. Vijay, S. K. Rajesh, and R. Prasad, "Biogas scrubbing, compression and storage: perspective and prospectus in Indian context," Renewable Energy, vol. 30, no. 8, pp. 1195–1202, 2005.

6. G. Nonhebel, Gas Purification Processes, George Newness, London, UK, 1964.

7. J. Cebula, "Biogas purification by sorption techniques," Architecture Civil Engineering Environment, vol. 2, pp. 95–103, 2009.

8. A. Petersson and A. Wellinger, "Biogas upgrading technologies—developments and innovations," IEA Bioenergy, Task 37-Energy from biogas and landfill gas, 2009, http://www.iea-biogas.net.

9. K. Ken, A. Don, J. P. Batmale, B. John, R. Brad, and S. Dara, "Biomethane from dairy waste-a sourcebook for the production and use of renewable natural gas in California," 2005.

10. R. Chandra, V. K. Vijay, P. M. V. Subbarao, and T. K. Khura, "Performance evaluation of a constant speed IC engine on CNG, methane enriched biogas and biogas," Applied Energy, vol. 88, no. 11, pp. 3969–3977, 2011.

Purification of Natural Gas with High Co$_2$ Content by Formation of Gas Hydrates: Thermodynamic Verification

N. Azmi, H. Mukhtar, and K.M. Sabil

Department of Chemical Engineering, Universiti Teknologi PETRONAS, 31750, Tronoh, Malaysia

ABSTRACT

High carbon dioxide (CO_2) content in natural gas may constitute some environmental hazards when release to the atmosphere. A variety of conventional separation methods are presently being used to remove the undesired gas fraction from crude natural gas. One promising approach to capture CO_2 from natural gas is by formation of gas hydrate. Gas hydrates can be formed in a system containing water and small molecule gases such as CH_4 and CO_2 at appropriate pressure and temperature conditions. It is important to gain accurate data of the phase behavior of the gas hydrate forming systems to ensure that the process conditions are set in hydrate forming conditions. In this study, thermodynamics modeling approached is implemented to generate the phase equilibria data since the phase behavior measurements are often expensive, tedious and time consuming processes. The thermodynamic program, CSMGem is successfully used for prediction of equilibrium conditions for single and binary hydrate former systems with AAD% is less than 10%. The program is being further used to predict gas hydrate equilibrium for natural gas with different concentration of CO_2.

INTRODUCTION

Nowadays, natural gas has become an important source of energy and feedstock for chemical industries (Scholz et al., 1981; Xiao et al., 2009). Natural gas is a mixture of combustible gases formed underground by the decomposition of organic materials in plant and animal. Raw natural gas is composed of several gases. The main component is methane and it also contains varying amounts of heavier hydrocarbons, acid gases, water, mercury and inert gases (Mokhatab et al., 2006).

Malaysia has the largest natural gas reserved among the Southeast Asian economies and is the third largest amongst the Asia Pacific economies. As per 1st January 2000, the recoverable reserves of Malaysian natural gas stand at 84.4 trillion standard cubic feet of which 48% is located at offshore Sarawak, 43% offshore east coast of Peninsular Malaysia and the remaining 9% at offshore Sabah. These large gas reserves are sufficient to last around 43 years with current production rate (Zulkifli et al., 2002).

In natural gas, non-hydrocarbon gases (CO_2, N_2, H_2S) can account between 1 to 99% of overall composition (Thrasher and Fleet, 1995). High CO_2 concentrations are encountered in diverse areas including South China Sea, Gulf of Thailand, Central European Pannonian basin, Australian Cooper-Eromanga basin, Colombian Putumayo basin, Ibleo platform, Sicily, Taranaki basin, New Zealand and North Sea South Viking Graben (Thrasher and Fleet, 1995). The composition of CO_2 can reach as high as 80% in certain natural gas wells such as wells at the LaBarge reservoir in western Wyoming and the Natuna production field in Indonesia (Holder et al., 1988).

Due to stringent regulation on CO_2 content in commercial natural gas, high CO_2 content in natural gas has to be removed. Various methods for removing CO_2 have been suggested such as cryogenic fractionation, selective adsorption, gas absorption and membrane process. Although some of these processes have proved successful for the selective removal of CO_2 from multi-component gaseous streams, they still have some critical problems associated with large energy consumption, corrosion, foaminess and low capacity (Kang and Lee, 2000). Moreover, the current technologies cannot purify CO_2 effectively when CO_2 increase to 50-80% in the natural gas stream. Hence, new separation technology which is environmental friendly and with low operational cost must be developed to cater for the separation of CO_2 from this high CO_2 content natural gas. One promising approach to capture CO_2 from natural gas is through gas hydrate formation. When gas hydrates are formed from natural gas, the concentrations of natural gas components in hydrate phase are different than that in the original gas mixtures.

Gas or clathrate hydrates are ice-like crystalline compounds which are formed through combination of water and small guest molecules like CH_4, CO_2, etc. under suitable conditions of low temperature and high pressure (Sloan and Koh, 2008). In a gas hydrate molecule, water forms special cavities and guest molecules are trapped inside the cavities. Depending on the type and size of guest molecule presents, different gas hydrate structures can be formed.

The three most common types of clathrate hydrate: structure I (sI), structure II (sII) and structure H (sH) have been well defined by Sloan and Koh, 2008. The type of hydrate that forms will highly depend on the composition of the gases in the feed as well as temperature and pressure of the system.

Over the last decade, the interest in using clathrate hydrates formation as separation method or storage and transportation medium has revived, especially for natural gas and CO_2. In literature, separation of CO_2 from gas streams and its sequestration in geological formation by gas hydrate formation have been widely studied by Sabil et al. (2010a,b). When gas hydrates are formed from natural gas, the concentrations of natural gas components in hydrate phase are different than that in the original gas mixtures. For example, in the case of methane-carbon dioxide (CO_2-CH_4) mixture, the hydrate phase will be richer in CO_2 than that CH_4 at certain condition (Kang and Lee, 2000). This selective information is the basis for utilization of gas hydrate formation as a separation process.

In order to successfully implement the hydrate formation as a method for separation of CO_2 from natural gas stream, the phase equilibria data need to be determined. The phase boundaries will limit the region in which this technology can be used for the separation process. In this study, thermodynamics modeling approached is implemented to generate the phase equilibria data since the phase behavior measurements are often expensive, tedious and time consuming processes. Since the modeling approached has been selected, a verification of the model is initially with some available literature data for single hydrate former system. The AAD% is calculated between the experimental and the modeling results. Once the model is proven suitable, the model is used to generate data for the binary and multi components systems.

THERMODYNAMIC APPROACH

The pressure and temperature conditions for the formation or dissociation of gas hydrate are governed by the thermodynamic equilibrium (Sloan and Koh, 2008). A system is in thermodynamic equilibrium when it is in thermal, mechanical and chemical equilibrium. For a system at constant pressure and temperature, thermodynamic equilibrium can be characterized by the minimization of Gibbs energy. According to standard thermodynamic phase equilibrium criteria, the chemical potential of each component must be the same in every phase at equilibrium conditions.

$$\mu_A^1 = \mu_A^2 = ... = \mu_A^k \tag{1}$$

Where, μ_i^1 is the chemical potential of component A in phase 1 and k is the number of coexisting phases for multiphase multicomponent equlibria.

Based on the above equations, the equilibrium condition may be calculated either by direct minimization of the Gibbs' energy or by using the principle of equality of chemical potentials (Walas, 1985). The chemical potential can be expressed in terms of the fugacity of a component by the following Equation (2):

$$\mu = \mu^0 + RT \ln \frac{f(p)}{P^0}$$

(2)

Where μ^0 is the chemical potential at reference state, T is the temperature, R is the universal gas constant, P0 is the pressure at the reference state and f (p) is the fugacity as a function of pressure. Combination of Eq. 1 and 2 results in the equality of fugacities for the thermodynamic equilibrium under consideration:

$$f_A^1 = f_A^2, \; f_B^1 = f_B^2$$

(3)

Where, f is the fugacity of component A or B in phase 1 or 2.

The fugacity approach as proposed by Klauda and Sandler (2000) has been used to model the hydrate in equilibrium. Their approach is basically based on solving the condition of equal fugacities of water in the hydrate phase and the fluid phases as shown in Eq. 4.

$$f_W^H (T, P) = f_W^\pi (T, P, x)$$

(4)

The thermodynamic modeled has been developed in commercial available hydrate software, CSMGem. Ballard and Sloan (2004) have reported the schematic of development procedure for the selected hydrate program. It has been proven that the hydrate formation temperatures and pressures for uninhibited systems are predicted quite well by CSMGem as compared with four other commercial hydrate programs; CSMHyd, DBRHydrate, Multiflash and PVTsim (Ballard and Sloan, 2002).

The present study shows the gas hydrate boundary conditions for natural gas components with different compositions, pressure:

$$ADD(\%) = (1/N) \sum_{j}^{N_p} \left| \frac{P_{cal} - P_{exp}}{P_{exp}} \right|_j \times 100$$

(5)

and temperature conditions. To verify the accuracy of the model, absolute deviation (AAD%) has been calculated by comparing the predicted data with available literature data for single component systems such as methane, ethane, propane, carbon dioxide and nitrogen with water. Then, similar evaluations are carried out for binary gas systems with water.

RESULTS AND DISCUSSION

Equilibrium of Pure Hydrate Formers:

A few single components in natural gas were selected to predict the hydrate phase equilibria when they are in contact with water. Hydrate equilibrium data for single gas hydrates such as methane, ethane and propane will become a basis for further understanding if phase equlibria of water with binary hydrate former systems. Figure 1 and 2 show the predicted three phase equilibria (hydrate-liquid water-vapor), H-L$_w$-V for pure component systems; CH_4 and CO_2 with available measurement data. As shown in the figures, pressure increases steeply with increasing temperature. Methane hydrate system yield a good agreement with available literature data but the hydrate equilibrium curves for CO_2 hydrate is slightly fluctuating with literature data. Technically, methane with small size of molecules forms sI hydrate whereas intermediate size of carbon dioxide allows its molecules to occupy the large cavity ($5^{12}6^2$) of sI hydrate.

Table 1 shows the absolute average deviation (AAD%) for five different pure components that have been studied in this work. All the hydrate formers have less than 5% of AAD and it can conclude that the pressure prediction for pure components using CSMGem gives good agreement with the available literature data. A maximum deviation of 4% is obtained by propane as a hydrate former and followed by carbon dioxide. Such a graphical comparison of data and deviation calculation for single-component hydrate formers gave reassurance that the prediction for hydrate equilibrium are acceptable in order to proceed to binary and multicomponent predictions.

Figure 3 shows the predicted hydrate equilibrium data for five different pure components in natural gas. The hydrate equilibrium lines of all components in natural gas have been compared since the work is on the separation of CO_2 from natural gas.

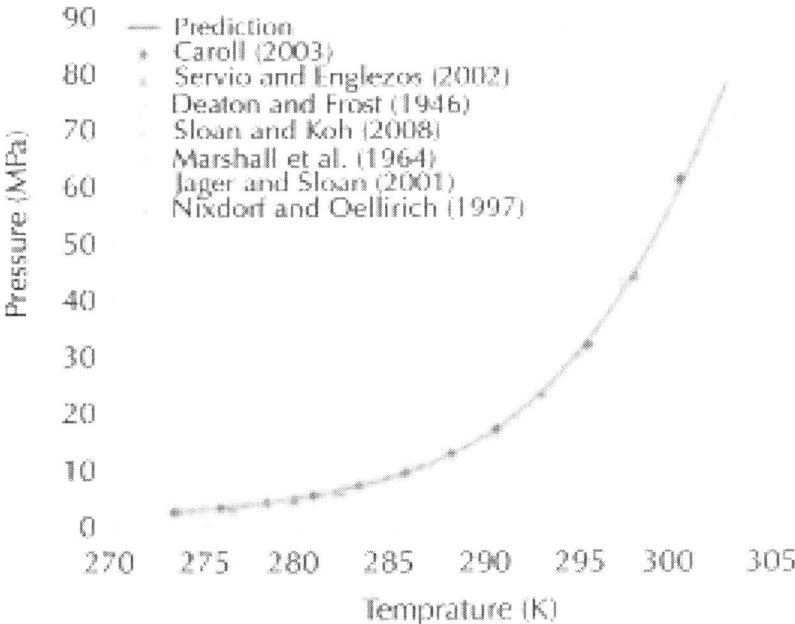

Figure 1: Three phase (H-Lw-V) equilibrium line for methane hydrate system.

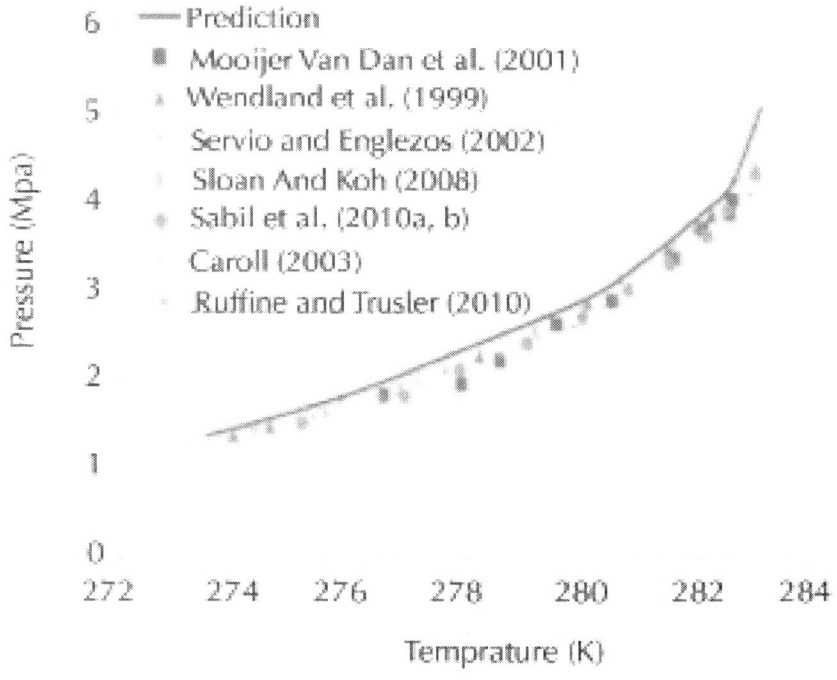

Figure 2: Three phase (H-Lw-V) equilibrium line for carbon dioxide hydrate system.

It been clearly observed that nitrogen tends to form a hydrate at higher pressure than the other four components at same temperature.

Due to differences in the volume and enthalpy of the vapour and liquid hydrocarbon, the three-phase hydrate formation line for ethane, propane and carbon dioxide change from H-L_w-V to H-L_w-L_{HC} (L_{HC} is liquid hydrocarbon). For each pure hydrate former, the predictions were bounded by the ice point (273 K) and the upper quadruple point (H-L_w-L_{HC}-V). Quadruple point is noted by the phases that are in equilibrium. Basically the lower quadruple point (Q_1) is approximately where the hydrate line intercepts the melting curve of pure water. Thus, all single hydrate formers have Q_1 approximately at 0°C (273 K). While the upper quadruple point (Q_2) is an interception within the hydrate line and the vapour pressure curve of the pure hydrate former.

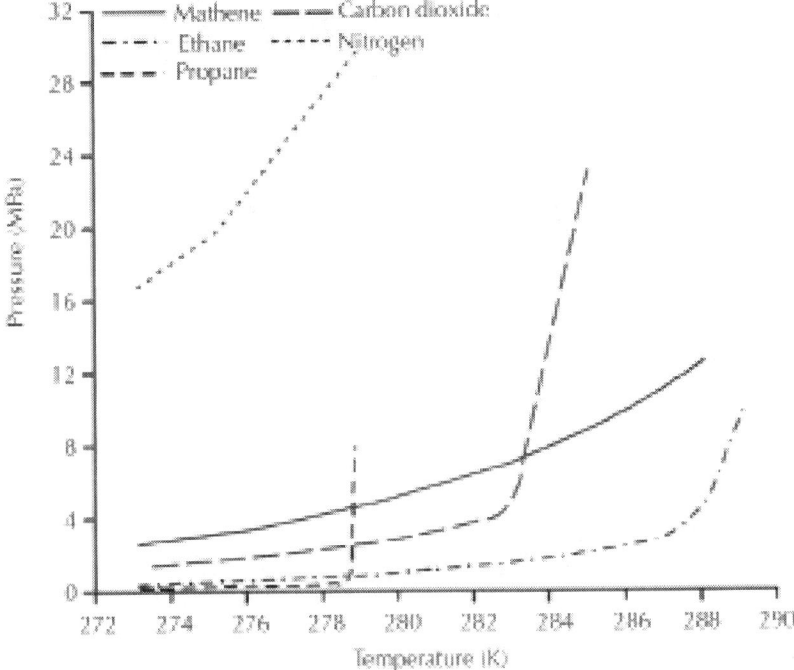

Figure 3: Predicted hydrate equilibrium of single components in natural gas.

Table 1: Absolute deviation (AAD %) results between the experimental data and predicted data for single hydrate systems

Hydrate former	No. of points	References	AAD (%)
Methane (CH4)	101	Servio and Englezos (2002)	1.5768
		Adisasmito et a (1991)	
		Jager and Sloan (2001)	
		Cairo! (2003)	
		Deaton and Frost (1940	
		Marshall et al. (1964)	
		Nixdorf and Oellrich (1997)	
Ethane (C2136)	62	Deaton and Frost (1946)	2.6357
		Nixdorf and Oellrich (1997)	
		Holder and Hand (1982)	

		Galloway et al. (1970)	
		Holder and Grigoriou (1980)	
		Avlonitis (1988)	
		Reamer et al. (1952)	
Propane (C3H8)	39	Deaton and Frost (1940	4.0267
		Nixdorf and Oellrich (1997)	
		Miller and Strong (1940	
		Robinson and Mehta (1971)	
		Kubota et al. (1984)	
Carbon Dioxide (CO2)	56	Adisasmito et al. (1991)	3.9056
		Cairo! (2003)	
		Wendland et aL (1999)	
		Mooijer-van den	
		Heuvel et al. (2001)	
		Servio and Englezos (2001)	
		Sabil et al. (2010b)	
		Ruffin and Trusler (2010)	
Nitrogen (N2)	73	Marshall et al. (1964)	1.8734
		Nixdorf and Oellrich (1997)	
		Van Cleeff and Diepen (1960)	
		Mohammadi et al. (2003) Jhaveri and Robinson (1965)	

Neither nitrogen nor methane has an upper quadruple since they have lower critical points which are far below the Q1. Such low critical temperatures prevent intersection of the vapour pressure line with H-L$_w$-V line above 273 K to produce an upper quadruple point. Methane and carbon dioxide hydrates have been compared in detail since methane is the main component in the natural gas. It can clearly been observed that carbon dioxide favor to form hydrate than that methane in the region T = 273.2 to 283.3 K and P = 1.2 to 7.4 Mpa. But the hydrate phase will start richer with methane above this region due to the sudden changes of hydrate equilibrium line of carbon dioxide hydrate from the upper quadruple point.

Equilibria of Binary Guest Mixtures:

Similar evaluations have been carried out for binary gas systems with water. Table 2 shows the outcomes of absolute deviation (AAD%) for binary mixtures of methane with ethane, propane, carbon dioxide and nitrogen.

Table 2 has shown that all the predicted data are in accordance with previously published work. In general, according to the Gibbs phase rule, for a ternary system (included water) which consisting of two gases + water, the three-phase equilibrium has two degrees of freedom. Thus, a second intensive variable need to be defined as an addition of temperature for equilibrium pressure prediction. The overall composition of the feed stream has been well defined for this case. There are a decrease of hydrate pressure in the mixtures of methane with ethane and propane as compared with pure methane hydrate. These phenomena happen due to the changing of hydrate structure which is from sI to sII. In contrast, the hydrate equilibrium pressure increases as concentration of nitrogen in methane+nitrogen system increase. As been discussed before, nitrogen molecules itself will form hydrate at higher pressure as compared with the other components. Thus, the presence of nitrogen in the mixture will increase the hydrate pressure until it reaches the pure nitrogen hydrate equilibrium line as the concentration of nitrogen in the system increase.

Basically it is not easy to generalize which hydrate structure will be present when sI and sII hydrate formers are in a mixture. In pure water, methane forms sI hydrate with its molecules occupying the small cavities (5^{12}). Since the molecular diameter of methane (4.4 Å) is smaller than the free diameters (~ 5.76 Å) of most cavities in the hydrate lattice, methane molecules can migrate in the hydrate lattice. Methane molecules will occupy the small cage in sII hydrate when there is the larger hydrate former present in the system. Thus, sII hydrate is formed for binary system of CH_4-C_3H8 since propane molecules cannot enter any of the sI cavities. In this system, methane will occupy the small cavities while propane molecules occupy the larger cavities of structure II hydrate. Although methane and ethane form sI hydrates by themselves, the mixture of these components form sII hydrates at certain compositions.

Table 2: Absolute deviation (AAD %) results between the experimental data and predicted

Mixtures	No. of points	References	AAD (%)
CH4-C2H6	59	Deaton and Frost (1946)	4.2517
		Nixdorf and Oellrich (1997)	
		Holder and Grigoriou (1980)	
CH4-C3H8	59	Deaton and Frost (1946)	5.3372
		Nixdorf and Oellrich (1997)	
		Verma et al. (1975)	
CH4-CO2	69	Adisasmito et al. (1991)	3.1032
		Zhu et al. (2005)	
		Fan and Guo (1999)	
CH4-N2	50	Nixdorf and Oellrich (1997)	8.4759
		Jhaveri and Robinson (1965)	
		Mei et al. (1996)	

Subramanian et al. (2000) have reported that the structure of methane+ethane hydrates changes from sI to sII over a methane vapor composition range (yCH_4) of 0.736–0.994 at 274.2 K.

The pressure versus temperature (P-T) diagram for CH_4-CO_2 mixtures have not been plotted due to inconsistency of CH_4 composition in the mixture from the literature data. But still the mixture has less AAD% as compared with the other three mixtures. Since the separation of CO_2 from natural gas mixture is the main objective of this work, P-T diagram for CH_4 - CO_2 system has been plotted using the predicted data (Fig. 4) at different concentration of CO_2. The purpose of doing this is to study the effect of CO_2 composition in the mixture. Hydrate equilibrium lines for pure CH_4 and CO_2 hydrates are plotted in the same graph for comparison purpose.

Phase equilibria of CH_4-CO_2 mixtures were investigated at temperature between 273.15 and 283.15 K. Mixture of CO_2 and CH_4 form sI hydrate only, like pure CO_2 and CH_4 with water (Uchida et al., 2005). In the mixture of 50% CH_4-50% CO_2, the equilibrium line is predicted close to CO_2 hydrate line instead of in the middle of the both single hydrate formers. Gaudette et al. (1996) have determined that the distribution coefficient of methane between the gas and hydrate phase is approximately 2 for mixture of 50/50 CH_4 - CO_2. Therefore, CO_2 hydrates do indeed form selectively over methane hydrates in the presence of 50/50 gas mixture. Seo et al. (2000) reported that the composition of CO_2 in the hydrate phase increase with increasing the CO_2 composition in vapor phase and decreasing the system pressure (Seo et al., 2000). This selective information will be the basis for utilization of gas hydrate formation as an approach to separate carbon dioxide from methane.

Equilibrium of Natural Gas with Different Concentration of Carbon Dioxide

Hydrate formation conditions were also been predicted for natural gas mixtures with and without CO_2. Several natural gases containing CO_2 from three gas production fields have been reported by Adisasmito and Sloan (1992) are listed in Table 3 (Adisasmito and Sloan, 1992). The pure CO_2 was included to indicate the whole range of concentration.

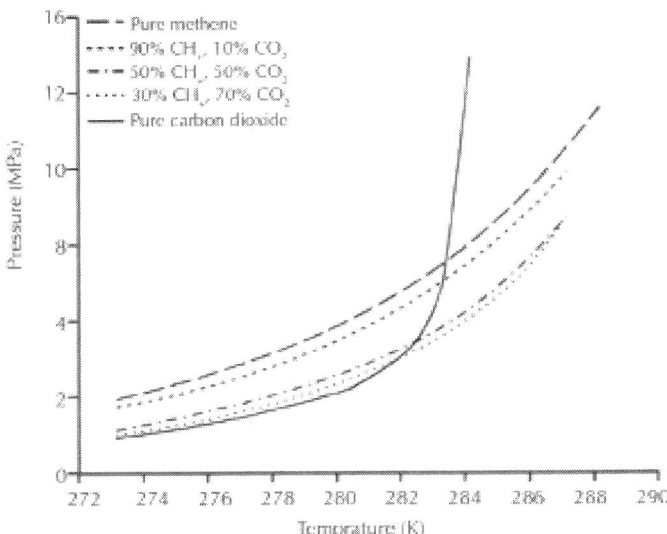

Figure 4: Predicted hydrate equilibrium data for CO_2-CH_4 system.

Table 3: Five different concentration of natural gas components

Component	Gas A	Gas B	Gas C	Gas D	Gas E	Gas F
CH4	76.62	52.55	24.42	12.38	7.86	-
C2H6	11.99	8.12	3.99	1.96	1.13	-
C3H8	6.91	4.74	3.07	1.66	0.86	-
i-C4H10	1.82	1.31	0.75	0.37	0.20	-
n-C4H10	2.66	1.88	0.92	0.48	0.33	-
CO2,		31.40	66.85	83.15	89.62	100000

The hydrate phase equilibria for natural gas system with overall carbon dioxide concentration of 31.4, 66.85, 83.15 and 89.62 mol% are depicted in Fig. 5. The temperatures have been further extended to predict the hydrate formation pressure since the temperature in literature is limited to 282 K. From the graph, it can be clearly observed that the equilibrium curve shift towards higher pressures as concentration of carbon dioxide in natural gas increase. The concentration of carbon dioxide in the system is obviously affecting the hydrate formation condition. The result obtained with the natural gases show the expected dependence of gas composition on the equilibrium conditions.

The hydrate equilibrium line rises vertically from the upper quadruple point (Q_2), with very large pressure changes for small temperature changes for each gas except gas A. Hydrate equilibrium line for gas A almost identical with that pure methane since the gas is mainly consist of methane. Structure II hydrates are formed in gas A, B and C which the small cavities are occupied by methane while the large cavities are mostly occupied by the other large components such as propane, isobutane and butane. In the mixtures have high concentration of CO_2 like gas D and E, structure I hydrates are formed where small cavities are occupied by methane. While the large cavities of structure I are being occupied by methane, ethane and carbon dioxide.

J. Applied Sci., 11 (21): 3547-3554, 2011

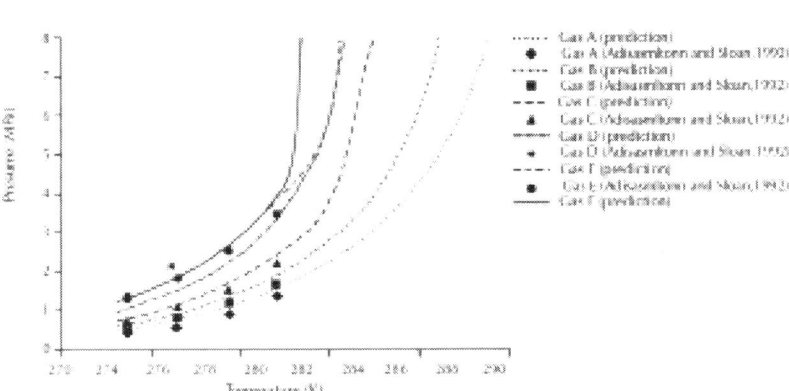

Figure 5: Hydrate equilibrium data of natural gas with different concentration of carbon dioxide.

In this condition, the very large molecules such as propane, butane and isobutane will act as gas diluents and do not participate in the structure I hydrate (Gaudette, et al., 1996). Therefore, both gases required higher pressure for structure I formation.

CONCLUSIONS

In this study, the thermodynamics model has been successful used to predict the phase equilibrium data for single, binary and multi components hydrate former systems. The predicted data of the model has been compared with experimental data for single hydrate former systems including CH4, CO_2 systems and the calculated AAD% is less than 5% for three phase equilibrium condition namely H-LW-V. The model is successful used to predict the mentioned equilibrium condition for binary and multi-component hydrate formers system. From this phase behavior data, the region where the hydrate formation can be used as separation process for CO_2 from natural gas can be identified.

ACKNOWLEDGMENTS

The authors are thankful to Universiti Teknologi PETRONAS for providing grant and facilities for the research purpose.

REFERENCES

1.   Adisasmito, S. and E.D. Sloan, 1992. Hydrates of hydrocarbon gases containing carbon dioxide. J. Chem. Eng. Data., 37: 343-349.

2.   Adisasmito, S., R.J. Frank and E.D. Sloan Jr., 1991. Hydrates of carbon dioxide and methane mixtures. J. Chem. Eng. Data, 36: 68-71.

3.   Avlonitis, D., 1988. Multiphase equilibria in oil-water hydrate forming systems. M.Sc. Thesis, Heriot-Watt University, Edinburgh, Scotland.

4.   Ballard, A.L. and E.D. Sloan Jr., 2002. The next generation of hydrate prediction: An overview. J. Supramol. Chem., 2: 385-392.

5.   Ballard, A.L. and E.D. Sloan Jr., 2004. The next generation of hydrate prediction IV: A comparison of available hydrate prediction programs. Fluid Phase Equilibria, 216: 257-270.

6.   Carrol, J., 2003. Natural gas hydrates-A guide for engineers. Gulf Professional Publishing, Imprint Amsterdam.

7.   Deaton, W.M. and E.M. Frost Jr., 1946. Gas hydrates and their relation to their operation of natural gas pipelines. Underground Storage. United States Department of the Interior-Bureau of Mines.http://www.prci.com/publicationsnew/L41020.cfm.

8.   Fan, S.S. and T.M. Guo, 1999. Hydrate formation of CO_2-Rich binary and quaternary gas mixtures in aqueous sodium chloride solution. J. Chem. Eng. Data, 44: 829-832.

9.   Galloway, T.J., W. Ruska, P.S. Chappelear and R. Kobayashi, 1970. Experimental measurements of hydrate numbers for methane and ethane and comparison with theoretical values. Ind. Enq. Chem. Fundam., 9: 237-243.

10. Gaudette, J., S. Al-Adel and P. Servio, 1996. Phase equilibria for the CO$_2$-CH4 mixed hydrate system. http://ppeppd07.chemeng. ntua.gr/manuscripts/51.pdf.

11. Holder, G.D. and G.C. Grigoriou, 1980. Hydrate dissocation pressure of (methane + ethane + water) existence of a locus of a minimum pressures. J. Chem. Thermodyn., 12: 1093-1104.

12. Holder, G.D. and J.H. Hand, 1982. Multiphase equilibria in hydrates from methane, ethane, propane, and water mixtures. AIChE J., 28: 440-447.

13. Holder, G.D., S.P. Zetts and N. Pradhan, 1988. Phase behavior in systems containing clathrate hydrates. Rev. Chem. Eng., 5: 1-70.

14. Jager, M.D. and E.D. Sloan, 2001. The effect of pressure on methane hydration in pure water and sodium chloride solutions. Fluid Phase Equilibria, 185: 89-99.

15. Jhaveri, J. and D.B. Robinson, 1965. Hydrates in the methane-nitrogen system. Can. J. Chem. Eng., 43: 75-78.

16. Kang, S.P. and H. Lee, 2000. Recovery of CO$_2$ from flue gas using gas hydrate: Thermodynamic verification through phase equilibrium measurements. Environ. Sci. Technol., 34: 4397-4400.

17. Klauda, J.B. and S.I. Sandler, 2000. A fugacity model for gas hydrate phase equilibria. Ind. Eng. Chem. Res., 39: 3377-3386.

18. Kubota, H., K. Shimizu, Y. Tanaka and T. Makita, 1984. Thermodynamic properties of R1$_3$ (CCIF$_3$), R$_{23}$ (CHF$_3$), R152a (C$_2$H$_4$F$_2$), and propane hydrates for desalination of sea water. J. Chem. Eng. Japan, 17: 423-429.

19. Marshall, D.R., S. Saito and R. Kobayashi, 1964. Hydrates at high pressures: Part I. Methane-water, argon-water, and nitrogen-water systems. AIChE J., 10: 202-205.

20. Mei, D.H., J. Liao, J.T. Yang and T.M. Guo, 1996. Experimental and modelling studies on the hydrate formation of a CH$_4$+N$_2$ gas mixture in the presence of aqueous electrolyte solution. Ind. Eng. Chem. Res., 35: 4342-4347.

21. Miller, B. and E.R. Strong, Jr, 1946. Hydrate storage of natural gas. Am. Gas Assoc. Monthly, 28: 63-67.

22. Mohammadi, A.H., B. Tohidi and R.W. Burgass, 2003. Equilibrium data and thermodynamics modelling of nitrogen, oxygen and air clathrate hydrates. J. Chem. Data, 48: 612-616.

23. Mokhatab, S., W.A. Poe and J.G. Speight, 2006. Handbook of Natural Gas Transmission and Processing. Chapter 1, Elsevier Inc., USA., ISBN-10: 0-7506-7776-7, pp: 672.

24. Mooijer-van den Heuvel, M.M., R., Witteman and C.J. Peters, 2001. Phase behaviour of gas hydrates of carbon dioxide in the presence of tetrahydropyran, cyclobutanone, cyclohexane and methylcyclohexane. Fluid Phase Equilibria, 182: 97-110.

25. Nixdorf, J. and L.R. Oellrich, 1997. Experimental determination of hydrate equilibrium conditions for pure gases, binary and ternary mixtures and natural gases. Fluid Phase Equilibria, 139: 325-333.

26. Reamer, H.H., F.T. Selleck and B.H. Sage, 1952. Some properties of mixed paraffinic and olefinic hydrates. AIME, 195: 197-205.

27. Robinson, D.B. and B.R. Mehta, 1971. Hydrates in the propane-carbon dioxide-water system. J. Can. Pet. Tech., 10: 33-35.

28. Ruffine, L. and J.P.M. Trusler, 2010. Phase behaviour of mixed-gas hydrate systems containing carbon dioxide. J. Chem. Thermodyn., 42: 605-611.

29. Sabil, K.M., G.J. Witkamp and C.J. Peters, 2010. Estimations of enthalpies of dissociation of simple and mixed carbon dioxide hydrates from phase equilibrium data. Fluid Phase Equilibria, 290: 109-114.

30. Sabil, K.M., G.J. Witkamp and C.J. Peters, 2010. Phase equilibria in ternary (carbon dioxide + tetrahydrofuran + water) system in hydrate-forming region: Effects of carbon dioxide concentration and the occurrence of pseudo-retrograde hydrate phenomenon. J. Chem. Thermodyn., 42: 8-16.

31. Scholz, W., G. Ranke, H. Becker, B.G. Bergo, A.I Grienko and A.V. Frolov, 1981. Method of Treating Natural Gas to Obtain a Methane Rich Fuel Gas. Vol. 4, US Patent Publication, USA., pp: 305-733.

32. Seo, Y.T., S.P. Kang, H. Lee, C.S. Lee and W.M. Sung, 2000. Hydrate phase equilibria for gas mixtures containing carbon dioxide: A proof-of-concept to carbon dioxide recovery from multicomponent gas stream. Korean J. Chem. Eng., 17: 659-667.

33. Servio, P. and P. Englezos, 2001. Effect of temperature and pressure on the solubility of carbon dioxide in water in the presence of gas hydrate. Fluid Phase Equilibria, 190: 127-134.

34. Servio, P. and P. Englezos, 2002. Measurement of dissolved methane in water equilibrium with its hydrate. J. Chem. Eng. Data, 47: 87-90.

35. Sloan, E.D. and C.A. Koh, 2008. Clathrate Hydrates of Natural Gases. 3rd Edn., CRC Press, Boca Raton.

36. Subramanian, S., A.L. Ballard, R.A. Kini, S.F. Dec and E.D. Sloan, 2000. Structural transitions in methane + ethane gas hydrates-Part I: Upper transition point and applications. Chem. Eng. Sci., 55: 5763-5771.

37. Thrasher, J. and A.J. Fleet, 1995. Predicting the Risk of Carbon Dioxide Pollution in Petroleum Reservoirs. In: Organic Geochemistry: Developments and Applications to Energy, Climate, Environment and Human History, Grimalt, J.O. and C. Dorronsoro (Eds.). AIGOA, San Sebastian, pp: 1086-1088.

38. Uchida, T., I.K. Ikeda, S. Takeya, Y. Kamata and R. Ohmura et al., 2005. Kinetics and stability of CH_4-CO_2 mixed gas hydrates during formation and Long-Term storage. ChemPhysChem, 6: 646-654.

39. Van Cleeff, A. and G.A.M. Diepen, 1960. Gas hydrates of nitrogen and oxygen. Rec. Trav. Chim., 79: 582-586.

40. Van der Waals, J.H. and J.C. Platteuw, 1959. Clathrate solutions. Adv. Chem. Phys., 2: 1-57.

41. Verma, V.K., J.H. Hand and D.L. Katz, 1975. Gas Hydrates from Liquid Hydrocarbons Methane-Propane-Water System. AIChe-VTG Joint Meeting, Munich, pp: 106.

42. Walas, S.M., 1985. Phase Equilibria in Chemical Engineering. Butterworth-Heinemann, Oxford,ISBN: 978-0750693134.

43. Wendland, M., H. Hasse and G. Maurer, 1999. Experimental pressure-temperature data on three-and four-phase equilibria of fluid hydrate, and ice phase in the system carbon dioxide-water. J. Chem. Data, 44: 901-906.

44. Xiao, Y., B.T. Low, S.S. Hosseini, T.S. Chung and D.R. Paul, 2009. The strategies of molecular architecture and modification of polymide-based membranes for CO_2 removal from natural gas: A review. Progress Polymer Sci., 34: 5561-5580.

45. Zhu, T., B.P. McGrail, A.S. Kulkarni, M.D. White, H. Phale and D. Ogbe, 2005. Development of a thermodynamic model and

reservoir simulator for the CH_4, CO_2 and CH_4-CO_2 gas hydrate system. SPE Western Regional Meeting, March 30-April 01, 2005, Irvine, California, SPE 93976.

46. Zulkifli, A.M., Y. Zulkefli, M. Rahmat and A.K. Yasmin, 2002. Managing our environmental through the use of clean fuel. Gas Technology Centre (GASTEG), Faculty of Chemical Engineering and Natural Resources Engineering, Universiti Teknologi Malaysia, Malaysia.

Chapter 5

Hydrogen from Biomass Gas Steam Reforming for Low Temperature Fuel Cell: Energy and Exergy Analysis

A. Sordi[1], E. P. Silva[2, 3], L. F. Milanez[3], D. D. Lobkov[2, 3], and S. N. M. Souza[4]

[1]Federal Technology University, Paraná, Rua Alagoas 200, 86020-430, Londrina - PR, Brazil

[2]University of Campinas, UNICAMP, Physics Institute, Hydrogen Laboratory, Campinas - SP, Brazil

[3]University of Campinas, UNICAMP, Faculty of Mechanical Engineering, Department of Energy, Campinas - SP, Brazil

[4]State University of West Parana, UNIOESTE, 85806-300, Cascavel - PR, Brazil

ABSTRACT

This work presents a method to analyze hydrogen production by biomass gasification, as well as electric power generation in small scale fuel cells. The proposed methodology is the thermodynamic modeling of a reaction system for the conversion of methane and carbon monoxide (steam reforming), as well as the energy balance of gaseous flow purification in PSA (Pressure Swing Adsorption) is used with eight types of gasification gases in this study. The electric power is generated by electrochemical hydrogen conversion in fuel cell type PEMFC (Proton Exchange Membrane Fuel Cell). Energy and exergy analyses are applied to evaluate the performance of the system model. The simulation demonstrates that hydrogen production varies with the operation temperature of the reforming reactor and with the composition of the gas mixture. The maximum H_2 mole fraction (0.6-0.64 mol.mol^{-1}) and exergetic efficiency of 91- 92.5% for the reforming reactor are achieved when gas mixtures of higher quality such as: GGAS2, GGAS4 and GGAS5 are used. The use of those gas mixtures for electric power generation results in lower irreversibility and higher exergetic efficiency of 30-30.5%.

INTRODUCTION

The fuel cell is an electrochemical device that converts the fuel chemical energy into electrical energy directly. The main classification of fuel cells is by type of electrolyte used and operation temperature. PEMFC uses NAFION®membrane as the electrolyte and gaseous diffusion electrodes with platinum (Pt) as catalyst, and the operating temperature is under 80°C. Due to the low temperature, the PEMFC operates only with hydrogen of high purity (99.99%mol.mol^{-1}), and the concentration of carbon monoxide in the gaseous flux should not exceed 10 μmol.mol^{-1}.

The main sources for hydrogen production are: water, fossil hydrocarbons and biomass. Technologies utilized to remove the hydrogen of those sources are: water electrolysis and hydrocarbon reforming.

Hydrogen can be obtained from biomass mainly through two processes, both involving the reforming of methane. The first is the anaerobic degradation of organic matter, producing biogas mainly composed of methane (CH_4) and carbon dioxide (CO_2). The second is biomass gasification, a thermo-chemical process in which biomass is transformed into fuel gas, usually called producer gas or synthesis gas (GGAS).

The gasification gas of biomass is composed of H_2, CO, CH_4, C_xH_y, CO_2 and N_2. The molar fractions of these gaseous species in the GGAS composition depend on the design of the gasifier, the biomass composition and operational conditions. The potential of gasification gas to produce hydrogen depends, in turn, on its composition, mainly on its CH_4 and CO contents. For electricity generation in a low-temperature fuel cell, such as the PEMFC, it is necessary a gas processing (reforming and purification) until there is only hydrogen in the output gaseous flow.

This work presents a methodology for the thermodynamic simulation of hydrogen production by reforming the gasification gas of biomass. Gas mixtures from different gasification processes were used in order to find which had the highest potential for hydrogen and electricity production. The "highest potential" mixture is understood as the one in which the reforming and purification of the gasification gas of biomass results in the minimum irreversibility for the total system.

PEMFC / GASIFICATION GAS SYSTEM

Figure 1 illustrates the model of the system for hydrogen and electricity production from gasification gas of biomass. The gas enters the reforming reactor at 850°C and atmospheric pressure at position 1, and leaves at position 2 at a temperature that depends on its composition at the entrance. The reform gas must be cooled before passing through the shift reactor for the reaction of carbon monoxide with steam. After the shift reactor, the shift gas must be cooled and compressed. The last part of the system is hydrogen purification by a PSA cycle and electrical energy production in PEMFC stacks. Currently, the PEMFC stacks are built on a power scale ranging up to 250 kW. In this work the fuel cell power was calculated based on the gas flow of GGAS1 pilot plant. Position 7 represents the flow of pure hydrogen, and position 9 represents the flow of the exhaust gas mixture at the PSA exit.

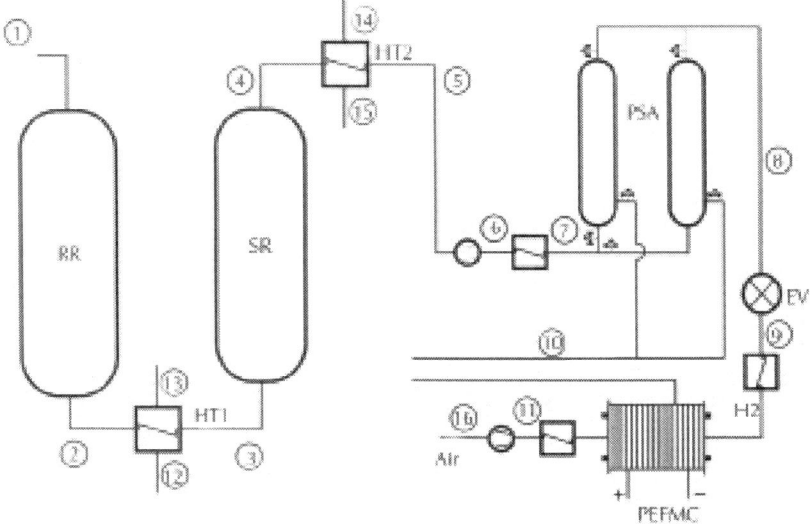

Figure 1: PEMFC / Gasification gas system.

THERMODYNAMICS OF THE FUEL CELL

According to Larminie and Dicks (2003), the thermodynamics of the fuel cell is represented by the reaction (1) and equations below. The Global reaction of hydrogen with oxygen from air is:

$$H_{2(g)} + \tfrac{1}{2}O_{2(g)} \rightarrow H_2O_{(l)}$$

Rc. (1)

The maximum electrical work W_{el} that can be produced by the fuel cell is:

$$-\delta W_{el} = dH - TdS$$

Or:

$$-W_{el} = \Delta G = \Delta H - T\Delta S \tag{1}$$

Eq. (1) can be written on a molar basis as:

$$\Lambda \overline{g} = \Delta \overline{h} - T \cdot \Delta \overline{s} \tag{2}$$

The available thermal energy is $\Delta \overline{h}$, therefore the maximum thermodynamic efficiency of the fuel cell is defined as:

$$\eta_{TH} = \frac{\Delta \overline{g}}{\Delta \overline{h}} \tag{3}$$

In reaction (1) two electrons are transferred to a circuit external to the fuel cell. For each mol of hydrogen involved in the reaction, 2N electrons are transferred, where N is the Avogadro's number. If -e is the charge of an electron then the charge flow will be (DOE, 2002):

$$-2N \cdot e = -2F \tag{4}$$

During the operation of the fuel cell, a difference of potential E appears between the cell electrodes. The electrical work w_{el} is calculated by the equation:

$$w_{el} = -2F \cdot E \tag{5}$$

If there are no irreversibilities present in the system, then the produced work will be equal to the Gibbs free energy of the reaction:

$$\Delta \bar{g} = -2F \cdot E \quad \text{and}$$

$$E = \frac{\Delta \bar{g}}{2F}$$

(6)

Equation (6) calculates the ideal voltage or reversible potential for a fuel cell operating with hydrogen. For standard conditions (298.15 K and 101.325 kPa), the following applies:

$$E^0 = \frac{-\Delta \bar{g}^0}{2F}$$

(7)

This results in 1.229 V for a fuel cell based in Rc.(1). However, several phenomena related to the kinetics of the electrochemical conversion in the electrodes result in losses (overpotentials) in the cell potential when the current intensity increases (Matelli and Bazzo, 2005). The overpotentials are known as: polarization by activation, ohmic polarization and polarization by concentration. The electrochemical efficiency quantifies the overpotential effects by the equation:

$$\eta_{elq} = \frac{E_{OP}}{E^0}$$

(8)

Where E_{OP} is the real operating voltage of the fuel cell. According to Matelli and Bazzo (2005), the practical efficiency can be defined by the thermodynamic efficiency multiplied by the electrochemical efficiency:

$$\eta_{PRT} = \eta_{TH} \cdot \eta_{elq}$$

(9)

For a fuel cell system operating with hydrogen from biomass gasification gas reforming, the overall efficiency is associated to the total power output and the gas lower heating value (LHV). This efficiency is named first law efficiency of the PEMFC/gasification gas system:

$$\eta_{LFC} = \frac{\dot{W}_{el}}{\dot{n}_{GGAS} \cdot LHV} = \frac{\dot{W}_{el}}{\dot{Q}_{GGAS}}$$

(10)

Where the electrical power of fuel cell is:

$$\dot{W}_{el} = \left(i \cdot 10^{-3} \cdot A\right) \cdot E_{OP}$$

(11)

GASIFICATION GAS REFORMING AND PURIFICATION

The gasification gas of biomass has a hydrogen molar fraction that can be maximized by the reforming of the molar fractions of methane (CH_4), hydrocarbons ($CxHy$), and carbon monoxide (CO).

For maximum conversion of these gaseous species into hydrogen two reactors are needed: one for hydrocarbon reforming and another for CO processing.

The process of reforming is defined as a thermochemical and catalytic conversion of a liquid, solid or gaseous fuel into a hydrogen-rich mixture. According to Silva (1991), most processes use light hydrocarbons for extracting hydrogen. Light hydrocarbons are those with molecular mass between methane and naphtha, and a boiling point below 250°C. These compounds can react with water at temperatures of 650-900°C. In the case of methane, a nickel/alumina catalyst is used.

The best known reforming methods are: steam reforming, partial oxidation, and auto-thermal reforming. In this work, steam reforming was considered. The global gasification gas reforming reactions are described by Rcs. (2) to (4).

$$CH_4 + aH_2O \rightarrow bCO + cCO_2 + dH_2 + eH_2O$$

Rc.(2)

$$C_2H_4 + fH_2O \rightarrow gCO + hCO_2 + iH_2 + jH_2O$$

Rc.(3)

$$CO + lH_2O \rightarrow mCO_2 + nH_2$$

Rc.(4)

The application of H_2 for power generation in PEMFC requires that the anode inlet gas have a CO concentration lower than 10 μmol.mol^{-1}, since CO is a poison to the fuel cell eletrocatalyst (Zalc and Löffler, 2002). If hydrogen is produced from hydrocarbons reforming (i.e. biomass gasification gas), purification is required in order to reduce the CO levels to cell requirements. This task is partially accomplished by a water gas shift (WGS) reactor (reaction 4) (Giunta et al., 2006). The final CO cleanup occurs in a preferential oxidation (PrOx) or in the pressure swing adsorption (PSA) unit.

The WGS reaction is moderately exothermic with a heat reaction, H= -40kJ.mol^{-1}. The high temperatures favor intrinsic kinetics while lower temperatures favor high equilibrium CO conversion.

Adiabatic operation with an inlet-outlet temperatures of 250-350°C yields poor performance because the temperature of the process stream increases to a point at which equilibrium conversion is low (<80%). Isothermal operation with inlet-outlet temperature of 250°C initially yields conversions lower than those obtained adiabatically, but the conversion curve continues to increase to an equilibrium conversion value of 90% (Zalc and Löffler, 2002).

Significantly better performance can be achieved by operating at a relatively high temperature and exploiting reaction kinetics when the gas composition is far from equilibrium and then lowering the temperature as thermodynamics begins to limit the CO conversion (Zalc and Löffler, 2002). This task is accomplished in two adiabatic stages using two different catalysts with intermediate cooling. The first reactor operates between 300 and 500°C (high temperature stage HTS) and uses an iron-based catalyst (Fe/Cr). The second reactor operates at lower temperatures (180-300ºC) (low temperature stage LTS) and uses a copper-zinc catalyst supported on alumina Cu/Zn/Al (Francesconi et al., 2007).

Hydrogen final cleanup can be achieved by a PSA system, which is widely used for gas purification. PSA is generally employed for oxygen or nitrogen from air, and for hydrogen generated by processes such as hydrocarbon reforming. This technology has been in commercial use for hydrogen purification since 1966, and is currently widely used (Myers et al., 2002).

Basically, a PSA works by the action of an adsorbent bed selective for certain gaseous species. A gaseous mixture is introduced into the bed under high pressure, and the adsorbing solid selectively adsorbs certain components, allowing the non-adsorbed component to pass through the bed as a purified gas.

The PSA systems operate in cycles, where three steps are basic to any process: pressurization, adsorption and depressurization. In the depressurization occurs the adsorbent regeneration and desorption of components retained, and then the process returns to its initial condition. Therefore, the removal of species adsorbed is done by total pressure reduction, which gives the PSA systems a faster pace in cycles and greater production per unit volume of adsorbent bed than other types of adsorption processes (Neves and Schvartzman, 2005). The fastness and the operation with two or more sync beds allows input and output of products continuously, these features are essential for fuel cell systems.

The energy required for this separation of gaseous species is obtained from the mechanical work of compressing the gaseous mixture. Energy expended in this mechanical work is a significant component of the operational cost of a PSA system.

SYSTEM PERFORMANCE ANALYSIS

The methodology used to analyze the performance of the system was the energy and exergy balance. Initially the simulation of the gasification gas reforming was carried out, in which the final mixture composition, corresponding to the chemical equilibrium in a given thermodynamic condition, was calculated using Lagrange multipliers to find the minimization point of the total Gibbs free energy of the system. In this work, the EES (Engineering Equation Solver) was used in the simulation of the gasification gas reforming.

The hydrogen molar fraction in the mixture is a function of the temperature, pressure and the steam/carbon ratio (γ). In the present work, the pressure was 101.3 kPa, and the parameter γ was equal to 2 for CH_4 reforming, 3 for C_2H_4, and 1 for CO reforming.

With regard to input conditions, the gasification gas may have compositions depending on the gasification process and biomass type. Table 1 illustrates the molar fractions of some gaseous mixtures produced by different gasification processes reported by Bain (2004). Initials in Table 1 stand for:

CFB: circulating fluidized bed

FB: fluidized bed

IFB: indirect fluidized bed

ICFB: indirect circulating fluidized bed

DRF: downdraft

UPF: updraft

Once the hydrogen molar fraction in the gaseous mixture in the shift reactor (shift gas) is calculated, the next step is the hydrogen purification by PSA. In the present work, a PSA system was considered with an operation pressure of 650 kPa, and a hydrogen recovery factor of 0.85 of the volumetric flow of the molar fraction of this species in the shift gas. These characteristics were chosen based on research and development data from the Hydrogen Laboratory of the University of Campinas Physics Institute.

Thermodynamic analysis of the system illustrated in Fig. 1 assumed the ideal gas model. Equations (12) and (13) represent the enthalpy and entropy calculations for an ideal gas mixture.

Table 1: GGAS composition for different gasification processes

Process	CFB	ICFB	FB	IFB	IFB	FB	DRF	UPF
Fluid reagent	air	stream	air	stream	stream	air	air	air

Biomass	Bagasse	Wood	Wood	Balck liquor	Wood	Wood	Wood	Wood
Mol. mol-1	GGAS1	GGAS2	GGAS3	GGAS4	GGAS5	GGAS6	GGAS7	GGAS8
YH2	10.0	26.2	21.7	29.4	31.5	11.0	16.0	10.0
YCO	12.7	38.2	23.8	39.2	22.7	17.0	21.5	14.8
YCO2	16.7	15.1	9.4	13.1	27.4	18.0	14.4	12.8
YN2	56.4	1.6	41.6	0.9	3.2	44.0	44.8	57.5
YCH4	3.7	14.9	2.9	13.0	11.2	7.0	3.3	4.9
YC2H4	0.5	4.0	0.6	4.4	4.0	3.0	0.0	0.0

$$\bar{h}_i = \sum_{i=1}^{n} y_i \cdot \left(\bar{h}_i^0 + \int_{T0}^{T} \bar{c}p_i \, dT \right)$$

(12)

$$\bar{s}_i = \sum_{i=1}^{n} y_i \cdot \left(\bar{s}_i^0 + \int_{T0}^{T} \left(\frac{\bar{c}p_i}{T} \right) dT - \bar{R} \ln(y_i) \right)$$

(13)

The mass and energy balances in reactors and heat exchangers are represented by Eqs. (14) to (16).

$$\sum \dot{m}_{in} = \sum \dot{m}_{out}$$

(14)

$$\sum_{i=1}^{n} \dot{n}_i \left(\bar{h}_i(T) \right)_P - \sum_{i=1}^{n} \dot{n}_i \left(\bar{h}_i(T) \right)_R = 0$$

(15)

$$\sum_{i=1}^{n} \dot{n}_{in} \cdot \bar{h}_{in} - \sum_{i=1}^{n} \dot{n}_{out} \cdot \bar{h}_{out} = \dot{Q}$$

(16)

The energy balance for a control volume is:

$$\dot{W}_{CV} - \dot{I} = \dot{n}_{out}\left(\overline{ex}_{out}\right) - \dot{n}_{in}\left(\overline{ex}_{in}\right)$$

(17)

The definition of specific physical and chemical energy, according to Szargut et al. (1988), is represented by Eqs.(18) and (19):

$$\overline{ex}_{PH} = \sum_{i=1}^{n}\left(\overline{h}_i - \overline{h}_i^0\right) - T_0\left(\overline{s}_i - \overline{s}_i^0\right)$$

(18)

$$\overline{ex}_{CH} = \sum_{i=1}^{n}y_i.\overline{ex}^0_{CHi} + \overline{R}.T_0.\sum_{i=1}^{n}y_i.\ln.y_i$$

(19)

Where the total specific energy is:

$$\overline{ex} = \overline{ex}_{CH} + \overline{ex}_{PH}$$

(20)

The reference environment proposed by Szargut et al. (1988), where standard temperature and pressure (298.15 K and 101.325 kPa) and the standard atmosphere composition can be found was used to define physical and chemical exergies.

Exergetic efficiency calculations adopted the input/output criterion defined by Kotas (1995) as ratio efficiency. The ratio efficiencies of the following control volumes were calculated: reforming and shift reactors, heat exchangers, PSA, PEMFC, and the system as a whole. Eqs.(21) to (26) represent these efficiencies, respectively.

$$\Psi_{,RR} = \frac{\dot{n}\left(\overline{ex}_{CH}(REFGAS)\right)}{\dot{n}\left(\overline{ex}_{CH}(H_2O) + \overline{ex}_{PH}(H_2O)\right) + }$$
$$\dot{n}\left(\overline{ex}_{CH}(GGAS) + \overline{ex}_{PH}(GGAS)\right)$$

$$(21)$$

$$\Psi_{,SR} = \frac{\dot{n}\left(\overline{ex}_{CH}(SHIFTGAS)\right)}{\dot{n}\left(\overline{ex}_{CH}(REFGAS) + \overline{ex}_{PH}(REFGAS)\right)}$$

$$(22)$$

$$\Psi_{,PSA} = \frac{\dot{n}\left(\overline{ex}_{PH}(H_2)\right)}{\dot{W}_{PSA}}$$

$$(23)$$

$$\Psi_{,HE} = \frac{\dot{n}\left(\overline{ex}_{PH,COLD,out} - \overline{ex}_{PH,COLD,in}\right)}{\dot{n}\left(\overline{ex}_{PH,HEAT,in} - \overline{ex}_{PH,HEAT,out}\right)}$$

$$(24)$$

$$\Psi_{,PEMFC} = \frac{\dot{W}_{el}}{\dot{n}\left(\overline{ex}_{CH}(H_2)\right) + \dot{W}_{AI}}$$

$$(25)$$

$$\eta_{II,FC} = \frac{\dot{W}_{el}}{\dot{n}\left(\overline{ex}_{CH}(GGAS)\right) + \dot{W}_{AI} + \dot{W}_{PSA}}$$

$$(26)$$

RESULTS

Figures 2 and 3 show the molar fractions of the equilibrium composition of GGAS1 and GGAS2 reforming simulations respectively. The potential for hydrogen production is higher from GGAS2 than from GGAS1 because of its higher CH_4 and CO molar fractions. The temperature at which hydrogen production reaches its peak depends on the amount of CH_4, CO and inert gases such as N_2 and CO_2. The CH_4 reforming reaction is enhanced by higher temperatures (650-850°C), while the CO reforming reaction is enhanced by lower temperatures (250°C) in an isothermal reactor as Zalc and Löffler (2002). Thus, gas cooling is necessary at the reforming reactor exit to allow the conversion of the remaining CO molar fraction in the shift reactor.

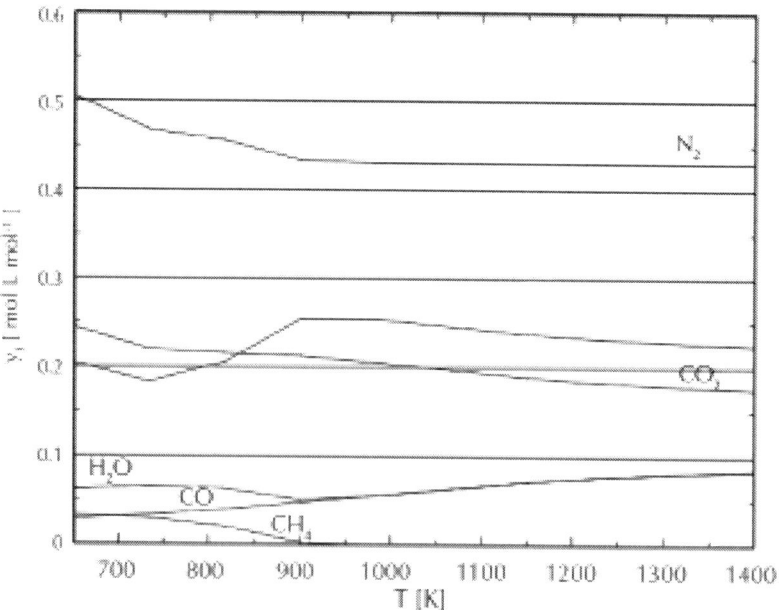

Figure 2: Molar fractions from GGAS1 reforming simulation.

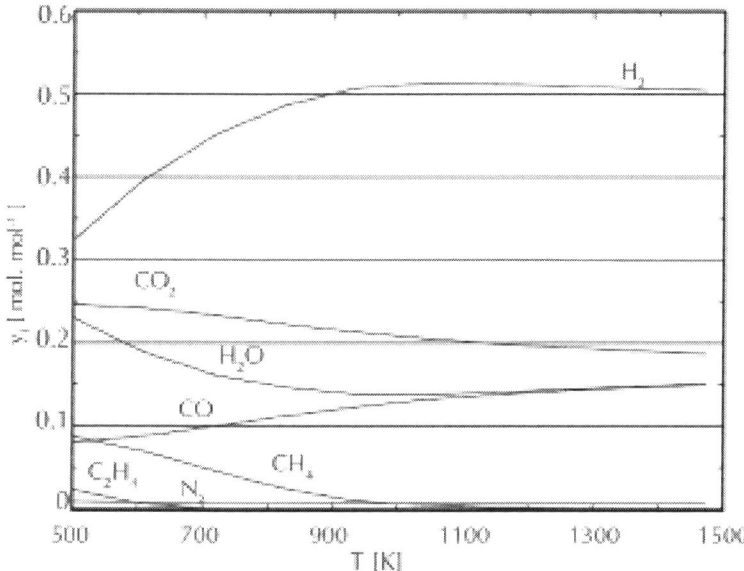

Figure 3: Molar fractions from GGAS2 reforming simulation.

In order to allow the use of the gasification gas physical energy at the reforming reactor, the gas pre-cleaning must be performed by the hot system. If the system is cold, the gaseous mixture must be heated up to the reforming operation temperature again, increasing the energy consumption of the hydrogen production system.

Tables 2 and 3 illustrate the results of thermodynamic simulation for PEMFC / GGAS system in Fig. 1.

Table 2: Simulation of PEMFC / GGAS1 system

QGGAS [kW]	ExCH [kW]	Wel [kW]	WAI [kW]	WPSA [kW]	I,FC [%]	II,FC [%]
1368.0	1371.0	549.8	71.1	119.8	40.2	26.2

Table 3: Simulation of PEMFC / GGAS1 system

Position	m	T	ex pH	eXcH	Ex
	kg/s	°C	(kJ/kg]	kj/kg	kW
1 (gas)	0.392	850	557.3	3,497.0	1,589.28
1 (water)	0.053	850	924.1	50.26	52.2
2	0.445	652	376.5	3,112.0	1,563.0
3	0.445	250	137.4	3,112.0	1,446.0
4	0.445	250	137.4	3,068.0	1,426,4
5	0.432	25	0.0	3,101.0	1,375.0
6	0.432	230.2	252.2	3,101.0	1,487.0
7	0.432	40	187.6	3,101.0	1,459.0
8	0.009	40	2,192.0	117,113.0	1,072.0
9	0.009	-	765.2	117,113.0	1,059.17
10	0.423	25	0.0	764,0	323 2
11	1.215	82.6	57.8	54.17	137.7
12	0.074	25	0.0	50.26	3.72
13	0.074	250	585.2	50.26	47.0
14	0.022	25	0.0	5026	1.1
15	0.022	200	545.5	50.26	13.12
16	0.001	25	0.0	117,113.0	147.4

Figure 4 shows that the largest contribution in the energy destruction (irreversibility) occurs in the PEMFC. The process of electrical work transfer in the fuel cell is associated to the chemical potential gradient, and the generation of irreversibility is proportional to this gradient. The exhaust gas temperature of the PEMFC is very close to the ambient temperature, and therefore there is not much heat produced by the PEMFC available.

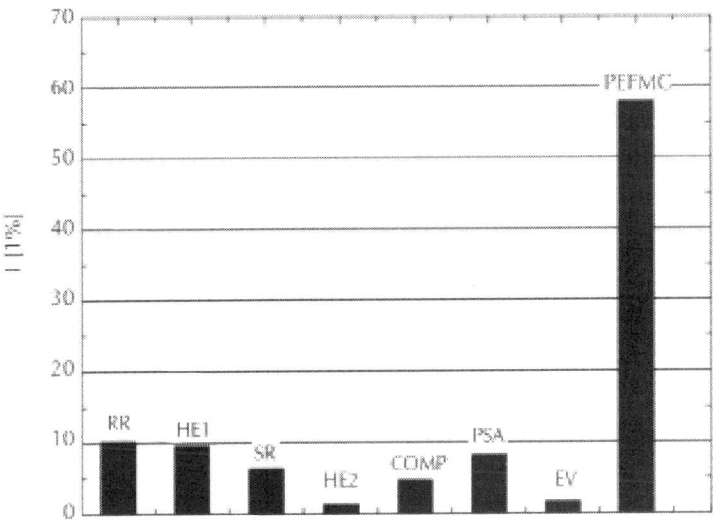

Figure 4: Percent Irreversibilities of PEMFC / GGAS1 system.

The gaseous mixture composition has an important effect on the system performance. For the gaseous mixtures in Table 1, Figure 5 shows the variation on reformer energetic efficiency as a function of the nitrogen molar fraction in the gasification gas of biomass. The higher the amount of this inert gas is, the lower the reformer efficiency. In practice, the presence of inert gases has a negative influence on the reaction kinetics.

Although the effectiveness of the reforming process is very important for the performance of the system, the PSA has the most striking influence on the energy consumption. Hydrogen purification is the main determinant of energy consumption, due to the need to compress the gaseous mixture, and due to the use of part of the purified hydrogen for adsorption bed regeneration. Figure 6 shows that purification performance is proportionally better for higher molar fractions of hydrogen in the shift gas. This favorable condition is obtained in the reforming of GGAS2, GGAS4 and GGAS5 (Figure 7). These gaseous mixtures are produced by the gasification process with steam injection and indirect heating, which yields gases of a higher quality. In a pilot plant, steam production should be provided for the gasification process; part of the required heat can be obtained by burning the PSA exhaust gas.

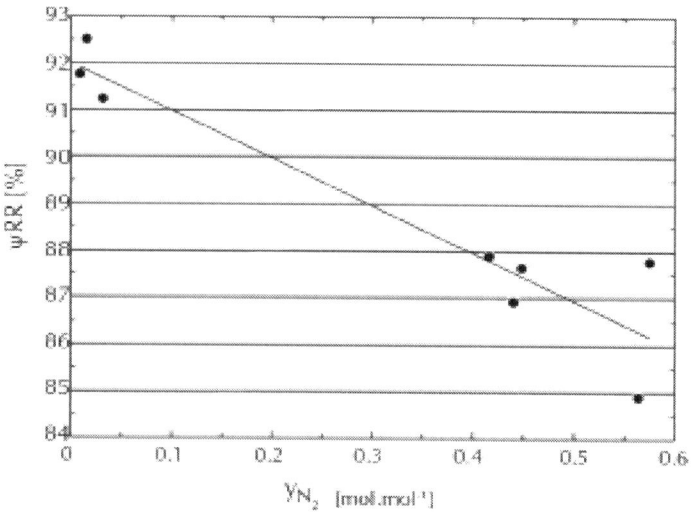

Figure 5: Exegetic efficiency of the reforming reactor as a function of nitrogen molar function.

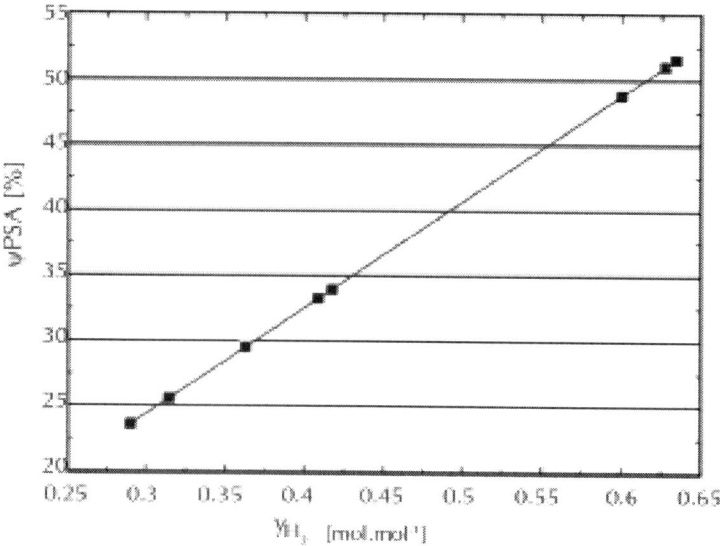

Figure 6: Exergetic efficiency of PSA as a function of hydrogen molar fraction in the SHIFTGAS.

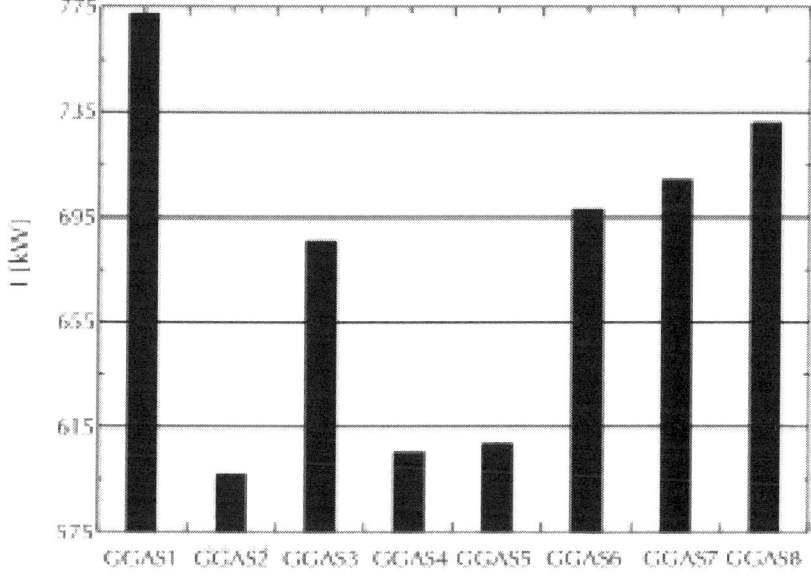

Figure 7: Total irreversibilities of system for different GGAS composition.

Figure 8 shows the variation of the first and second law efficiencies of the fuel cell system as a function of the hydrogen molar fraction in the shift gas. The first law efficiency is an adequate parameter to measure system performance but is not adequate to measure the gas quality, since the lower heating value does not reflect the quality of the gaseous mixture for hydrogen production. The second law efficiency, on the other hand, is indicative of the gasification gas quality in the sense that the smaller the gas flow required to produce a given amount of hydrogen, the higher the system performance will be. For the construction of a pilot plant, the higher exergetic efficiency illustrated in Figure 8 results in components of a smaller volume. Figure 9 also indicates that the system will be smaller due to the lower irreversibility. A pilot plant with smaller volume components will have a correspondingly lower cost.

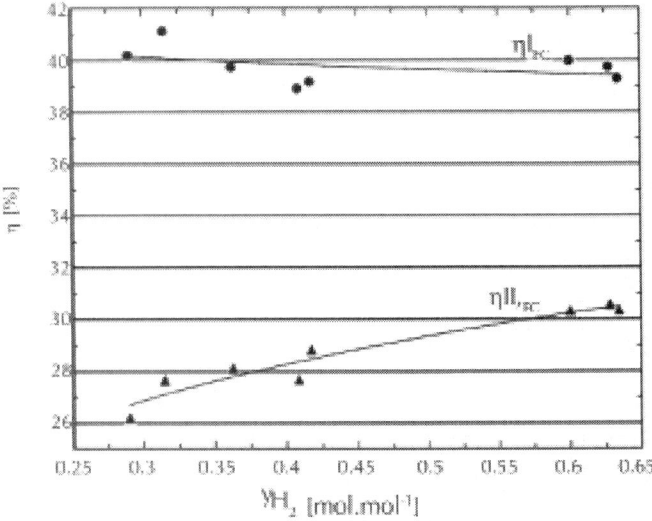

Figure 8: First and second law efficiencies as a function of the hydrogen molar fraction.

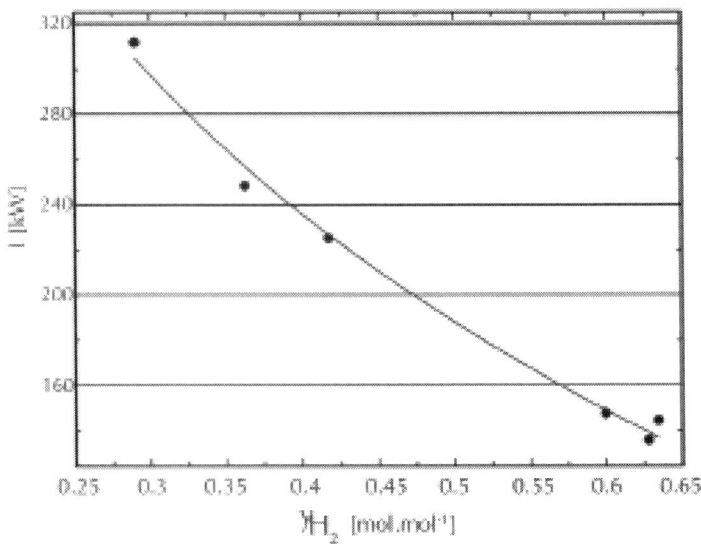

Figure 9: Total system irreversibility as a function of the H_2 molar fraction in the SHIFTGAS.

CONCLUSIONS

The simulation demonstrated that the hydrogen production varies with the operation temperature of the reforming reactor, and with the gas mixture composition. These two variables are inherently dependent because of the particular chemical equilibrium state for each thermo-chemical reaction of system species.

The temperature of 850°C in output gasification reactor is sufficient for maximum methane conversion and hydrogen production. But hot gas cleaning should be used in gasification process so that gasification gas physical energy can be utilized in steam reforming. In cold gas cleaning the cooling of the gasification gas to the environment temperature in the wet scrubber results in a thermal energy loss, which is a disadvantage of the cold cleaning. Gas reheating for the steam reforming reaction also causes the destruction of the energy inherent to the heat transfer process, and makes the system more complex.

The energetic efficiency of the reforming reactor also depends on the gas composition; for the lower quality gas (GGAS1) the efficiency was 85%. For GGAS2, which is the best quality gaseous mixture, the reactor efficiency presented a value of 92.5%.

Both the exergetic efficiency and the total irreversibility values of the power generation system with PEMFC were, respectively, 26.2% and 775 kW for GGAS1 and 30.5% and 575 kW for GGAS2. Therefore, to achieve the maximum performance of the hydrogen production and electricity generation in the fuel cell systems, the gasifier should be designed to obtain gas mixture similar to GGAS2.

The main responsible for the difference between the first and second law efficiencies of the PEMFC system is the hydrogen purification process (PSA). Thus, for a fuel cell / gasification gas system to be competitive when compared with the traditional thermal machines, further research and development in gas purification technology is necessary. On the other hand fuel cells operating at high temperatures like the SOFC (Solid Oxide Fuel Cell) and MCFC (Molten Carbonate Fuel Cell) are adequate to operate with gases from gasification because there is no need for hydrogen purification.

ACKNOWLEDGEMENTS

We are thankful to CAPES for the financial support of this work, and to the Hydrogen Laboratory and the Mechanical Engineering Faculty at UNICAMP.

REFERENCES

1. Bain, R. L., Biomass Gasification Overview, NREL National Renewable Energy Laboratory. US DOE United States Department of Energy, p. 48 (2004).]

2. DOE - U.S. Department of Energy. Fuel Cell Handbook, EG & G Services Parsons, Inc. Science Applications International Corporation, West Virginia, USA, p. 352 (2002).

3. Francesconi, J. A., Mussati, M. C., Aguirre, P. A., Analysis of design variables for water-gas-shift reactors by model-based optimization, Journal of Power Sources, v. 173, p. 467-477 (2007).

4. Giunta, P., Amadeo, N. and Laborde, M., Simulation of a Low Temperature Water Gas Shift Reactor Using the Heterogeneous Model/Application to a PEM Fuel Cell, Journal of Power Sources, v. 156 (7) p. 489-496 (2006).

5. Kotas, T. J., The Exergy Method of Thermal Plant Analysis. Krieger Publishing Company, Florida (1995).

6. Larminie, J. and Dicks, A. Fuel Cell Systems Explained, John Wiley Sons (2003).

7. Matelli, J. A., Bazzo, E., A Methodology for Thermodynamic Simulation of High Temperature Internal Reforming Fuel Cell Systems, Journal of Power Sources, v.142, p. 160-168 (2005).

8. Myers, B. D., Ariff, G. D., James, B. D., Lettow, J. S., Thomas, C. E. and Kuhn, R. C., Cost and Performance Comparison of Stationary Hydrogen Fueling Appliances, The Hydrogen Program Office; Office of Power Technologies; U.S Department of Energy. 123 p. (2002).

9. Neves, C. F. C and Schvatzman, M. M. A. M., Separação de CO_2 por meio da tecnologia PSA, Química Nova, v.28, n. 4, p. 622-628 (2005).

10. Silva, E. P., Introdução a Tecnologia e Economia do Hidrogênio. Campinas: Editora da Unicamp (1991).

11. Szargut, J., Morris, D. R. and Steward, F. R., Exergy Analysis of Thermal, Chemical and Metallurgical Processes, John Benjamins Publishing Co. (1988).

12. Zalc, J. M. and Löffler, D. G., Fuel processing for PEM fuel cells: transport and kinetic issues of system design, Journal of Power Sources, v. 111, p. 58-64 (2002).

chapter 6

Feasibility Study of an Alkaline-based Chemical Treatment for the Purification of Polyhydroxybutyrate Produced by a Mixed Enriched Culture

Yang Jiang[1], Gizela Mikova[2], Robbert Kleerebezem[1], Luuk AM van der Wielen[1], and Maria C Cuellar[1]

[1]Department of Biotechnology, Delft University of Technology, Julianalaan 67, Delft, 2628 BC, the Netherlands

[2]Polymer Technology Group Eindhoven BV, De Lismortel 31, Eindhoven, 5612 AR, the Netherlands

ABSRATCT

This study focused on investigating the feasibility of purifying polyhydroxybutyrate (PHB) from mixed culture biomass by alkaline-based chemical treatment. The PHB-containing biomass was enriched on acetate under non-sterile conditions. Alkaline treatment (0.2 M NaOH) together with surfactant SDS (0.2 w/v% SDS) could reach 99% purity, with more than 90% recovery. The lost PHB could be mostly attributed to PHB hydrolysis during the alkaline treatment. PHB hydrolysis could be moderated by increasing the crystallinity of the PHB granules, for example, by biomass pretreatment (e.g. freezing or lyophilization) or by effective cell lysis (e.g. adjusting alkali concentration). The suitability of the purified PHB by alkaline treatment for polymer applications was evaluated by molecular weight and thermal stability. A solvent based purification method was also performed for comparison purposes. As result, PHB produced by mixed enriched cultures was found suitable for thermoplastic applications when purified by the solvent method. While the alkaline method resulted in purity, recovery yield and molecular weight comparable to values reported in literature for PHB produced by pure cultures, it was found unsuitable for thermoplastic applications. Given the potential low cost and favorable environmental impact of this method, it is expected that PHB purified by alkaline method may be suitable for other non-thermal polymer applications, and as a platform chemical.

INTRODUCTION

Polyhydroxyalkanoates (PHAs) have received much attention as bio-based plastics that may contribute to future replacement of petroleum based plastics. Their performance ranges from stiff and brittle to soft and tough (Sudesh et al. [2000] and Laycock et al. [2013]). The most common PHA is polyhydroxybutyrate (PHB), which has similar thermal and some mechanical properties (e.g. tensile strength) compared to isotactic polypropylene (Sudesh et al. [2000]). In contrast to petroleum based plastics, PHA's biodegradability in various natural environments makes them suitable as disposables for packaging, agricultural or medical applications (Williams and Martin [2002], Bucci et al. [2005], Markets and Markets, [2013]). The fact that more and more varieties of

PHAs have been discovered and/or synthesized suggests that PHAs are not limited to thermoplastic applications. Moreover, PHA derivatives such as hydroxy fatty acid monomers may serve as chiral building blocks for the production of biochemicals and the methyl esters of their monomers could be used as a biofuel (Chen, [2009]).

Chen ([2009]) summarized the current status of commercial PHA production. Many types of commercial PHAs are available on the market. For example, Polyhydroxybutyrate-co-hydroxyvalerate (PHBV) can be synthesized by pure culture of either *Ralstonia eutropha* or recombinant *E. coli* from glucose and propionic acid. Middle chain length PHAs, such as polyhydroxyhydroxyhexanoate (PHHx), can be produced by pure culture of *Pseudomonas putida*. Despite of the above mentioned advantages of PHAs compared to conventional petroleum based plastics their large scale application is still constrained by their high price in the market. Economic evaluations of the PHA production process identified the following cost drivers (Choi and Lee,[1997]; van Wegen et al. [1998]): (a) raw materials (fermentation feedstock), (b) downstream processes for product recovery and purification, and (c) costs associated to maintaining a pure culture during the fermentation (e.g. fermentor costs and energy required for sterilization). Several studies have integrated the PHA production process with wastewater treatment with a dynamic feast-famine enrichment system, aiming at intracellular PHB content up to 90% (Johnson et al. [2009]), in order to reduce the cost from raw material and energy consumption aspects (reviewed by Dias et al. [2006]). Recent results showed that such process is capable of producing PHAs as good as the current pure-culture process in terms of intracellular PHAs content and biomass specific PHAs production rates (Jiang et al. [2012]). However, the challenge in terms of cost reduction in downstream process still remains.

PHAs are present in microorganisms as hydrophobic and water insoluble inclusion bodies which need to be separated from cell material. Plenty of techniques for PHA recovery and purification from pure cultures have been evaluated in literature and reviewed by Jacquel et al. ([2008]) and by Kunasundari and Sudesh ([2011]). The conventional organic solvent based purification method is still the best in terms of final product purity and recovery yield, although organic solvents may generate environmental issues (Ramsay et al. [1994]; de Koning and Witholt, [1997]). Several less toxic organic solvents have been reported for PHAs extraction (summarized in Jacquel et al.[2008]; Kunasundari and Sudesh [2011]; Riedel et al. [2013]). Most of those solvents are specific for middle chain length PHAs purification, instead of short chain length PHAs (e.g. PHB, PHV) (Jiang et al. [2006]; Elbahloul and Steinbüchel, [2009]; Terada and Marchessault [1999]). Nevertheless, short chain length PHAs are usually the main products when wastewater is used as feedstock (Dionisi et al. [2005]; Bengtsson et al. [2008]; Albuquerque et al. [2010]; Jiang et al. [2012]). Moreover, solvents such as 1, 2-proplene bicarbonate, require high temperature (>140°C) during the purification process, which typically leads to high energy consumption (Fiorese et al. [2009]; Riedel et al. [2013]).

Removal of cell materials by alkaline treatment was considered more economically feasible by Choi and Lee ([1997], [1999]) as compared to an organic solvent based PHA purification process. The alkaline treatment method has been widely reported in literature for pure cultures, resulting in purity and recovery yield as high as 98% and 97%, respectively (Choi and Lee [1999], Mohammadi et al. [2012a], [b]). An open culture process is based on the enrichment of a mixture of different microorganisms; it is unclear whether alkaline treatment can equally remove cell materials from microorganisms from enriched mixed cultures. Furthermore, the chemicals used in the treatment could degrade the PHA granules, as well as negatively influence the thermal stability of PHAs during processing as thermoplastics (Kim et al. [2006]).

The fate of PHAs during alkaline treatment and the thermal stability of the chemically treated PHA have hardly been reported. Moreover, only few studies have been published on recovery and purification of PHAs from mixed cultures (Serafim et al. [2008]). In this study, the feasibility of the alkaline method for recovery and purification of PHB obtained from mixed cultures was evaluated. This study focused on the PHA degradation during the chemical treatment and on product properties such as molecular weight and thermal stability. PHB recovery and purification by extraction with dichloromethane was used for comparison purposes.

MATERIAL AND METHODS

Biomass Preparation and PHB Recovery

The biomass used in this study was obtained from a 2 L sequencing batch reactor (SBR) fed with acetate under feast-famine condition. The composition of the working medium was: 125 mM NaAc $\cdot$ 3H$_2$O, 3.93 mM NH$_4$Cl, 1.87 mM KH$_2$PO$_4$, 0.42 mM MgSO$_4$ $\cdot$ 7H$_2$O, 0.54 mM KCl, 1.13 ml/L trace elements solution according to Vishniac and Santer ([1957]) and 3.71 mg/L allythiourea (to prevent nitrification). The operational conditions of the bioreactor were 30°C, pH 7, 1 day sludge retention time (SRT) and hydraulic retention time (HRT) and 18 h cycle length. The length of the feast phase was about 2.5 h during the steady state. The PHB was the sole storage polymer produced due to the fact that acetate was the sole carbon source. The biomass was collected at the end of the feast phase, when the cellular PHB content was between 62 wt% and 72 wt%. The dominant bacterial species in the SBR operated under such condition was *P. acidivorans*, a gram-negative bacterium (Jiang et al. [2011]).

Fresh biomass from bioreactor was collected by centrifugation (Heraeus, Germany) at 10000 g for 10 min at room temperature. The supernatant was removed and the pellet was resuspended with Milli-Q water to reach a final biomass concentration of approximately 20 g/L. 10 mL of this biomass suspension was used for PHB recovery. Two types of chemicals were applied either solely or together to remove the cell materials: (1) alkalis (NaOH at concentrations varying between 0.02 M and 1 M, or 0.2 M NH$_4$OH), and (2) surfactant (SDS at concentrations varying between 0.025% and 0.2%). The biomass suspension with the added chemicals was incubated in 50 mL tubes at 200 rpm and 30°C for 1 hour unless otherwise stated. The suspension was subsequently centrifuged at 10000 g for 10 min at 4°C. The supernatant was separated from the pellet and collected for soluble polymer or monomer measurements. The pellet was washed twice with Milli-Q water and dried at 60°C overnight.

Besides fresh biomass, pre-treated biomass was also evaluated in this study. The fresh biomass pellet collected after centrifugation was subjected to either freezing at −20°C or lyophilization. The same chemical treatment procedures as for the fresh biomass were applied to the pre-treated biomass in order to study the influence of pre-treatment on the PHB recovery. Lyophilized biomass was additionally used for solvent extraction for comparison purposes. The PHB was firstly purified by dichloromethane, following the procedure described in Ramsay et al. ([1994]). Further purification was achieved by dissolving 1 wt% of PHB in chloroform at 60°C for 50 min. The chloroform sample was subsequently slowly poured into cold ethanol (10 times volume to chloroform) while stirring rigorously. The precipitate was filtered of the solution, washed with ethanol and vacuum dried at 50°C.

The setup of all the experiments in this study is summarized in the Table 1. All experiments were performed in at least duplicate.

Table 1: Summary of all experiments conducted in this study

Chemical	Concentration	Biomass state	Time	Initial PHB content	Purity	Recovery	Mass balance	HB/PHBc
[−]	[M;w;v%]	[−]	[h]	[wt%]	[%]	[%]	[%]	[%]
NaOH	0.02	Fresh	1	71.8±5.7	77.3±4.0	92.2±4.2	−1.0±3.7	97.9±5.3
NaOH	0.05	Fresh	1	72.7±7.1	84.0±1.4	94.7±3.2	−3.9±2.9	96.7±4.2
NaOH	0.10	Fresh	1	65.3±3.1	83.8±4.4	98.0±1.4	−1.0±1.4	92.1±8.5
NaOH	0.20	Fresh	1	69.4±1.1	86.6±3.0	96.7±1.9	−0.7±2.6	97.5±16.7
NaOH	0.20	Fresh	0.3	68.6±0.7	87.3±2.2	96.4±2.6	−2.9±2.4	85.9±14.8
NaOH	0.20	Fresh	0.5	68.6±0.7	88.8±0.8	98.5±1.8	−0.7±1.8	90.9±18.9
NaOH	0.20	Fresh	3	68.6±0.7	92.1±0.8	93.5±2.4	−1.5±0.6	92.0±3.2
NaOH	0.40	Fresh	1	65.3±3.1	87.9±5.4	95.2±3.7	0.7±3.1	89.9±4.8
NaOH	0.70	Fresh	1	65.3±3.1	89.7±5.8	90.9±5.0	0.4±4.4	89.4±7.1
NaOH	1.00	Fresh	1	65.3±3.1	90.6±4.7	85.6±2.3	−0.3±2.1	89.1±4.0
NH4OH	0.20	Fresh	1	68.6±0.7	62.6±2.8	63.3±16.4	−3.9±0.9	36.3±10.9
SDS	0.20	Fresh	1	68.0±0.0	79.0±1.4	63.5±0.7	3.6±0.6	14.0±1.4
NaOH+SDS	0.20+0.025	Fresh	1	66.1±2.2	94.9±2.6	92.6±6.9	−2.7±3.8	94.2±6.4
NaOH+SDS	0.20+0.050	Fresh	1	66.1±2.2	96.9±1.3	93.5±4.8	−2.4±2.4	92.4±3.1
NaOH+SDS	0.20+0.100	Fresh	1	66.1±2.2	98.3±0.5	91.5±5.9	−3.9±4.8	96.3±4.8
NaOH+SDS	0.20+0.200	Fresh	1	66.1±2.2	99.1±0.5	91.0±4.9	−3.1±1.9	92.5±5.0
NaOH	0.20	Freezing	1	65.9±2.4	94.1±3.5	95.6±2.5	−2.9±2.1	94.3±5.4

NaOH	0.20	Freeze dried	1	69.9±2.2	95.9±3.7	95.5±0.6	-3.2±0.8	98.8±0.9
NH4OH	0.20	Freeze dried	1	69.9±2.2	87.4±2.1	95.0±1.8	-3.9±0.9	87.1±12.9
SDS	0.20	Freeze dried	1	69.9±2.2	93.5±4.1	93.7±2.2	-3.1±1.6	91.3±8.7
CH2Cl2	30a	Freeze dried	o/nb	72.2±0.4	97.6	55.9	ND	ND

[a]30 times of TSS.

[b]Overnight.

[c]Fraction of hydrolyzed monomer in total polymer in the supernatant.

Jiang et al.

Jiang et al. AMB Express 2015 5:5, doi:10.1186/s13568-015-0096-5

Analytical Methods

In order to evaluate the PHB mass balance of all experiments, the PHB quantity in fresh biomass, in final products and in the supernatant were determined. The PHB content in the biomass and in the final products was determined by gas chromatography (GC) according to the method described in Johnson et al. ([2009]). Commercial PHB (SigmaAldrich, the Netherlands) was used as standard. Based on the PHB mass present in the biomass (*PHAinitial*) and the dried pellet (*PHAend*), the recovery yield was calculated by equation 1:

$$Recovery = \frac{PHA_{end}}{PHA_{initial}} \cdot 100\% \qquad [g/g] \tag{1}$$

The PHB losses in the supernatant after chemical treatment (*PHB$_{supernatant}$*) was analyzed by gas chromatography (GC) with a modified procedure: 0.5 ml of the supernatant from chemical treatment was used for PHB concentration analysis. Commercial PHB mixed with 0.5 ml of chemical solution for PHB purification was used as standard. The remaining procedures were the same as described in Johnson et al. ([2009]). The potential by-products of chemical treatment (e.g. hydrobutyric acid, HB and crotonic acid, CA) (Yu et al. [2005]) were analyzed by high-performance liquid chromatography (HPLC) with a BioRad Animex HPX-87H column and a UV detector (Waters 484, 210 nm). The mobile phase, 1.5 mM H_3PO_4 in Milli-Q water, had a flow rate of 0.6 mL/min and a temperature of 59°C.

The overall mass balance was calculated by equation 2:

$$MassBalance = \frac{(PHA_{end} + PHB_{supernatant} + CA_{supernatant} - PHA_{initial})}{PHA_{initial}}$$
$$\times 100\% \qquad [g/g] \tag{2}$$

where, *PHB$_{supernatant}$* means the total PHB loss within the supernatant measured by GC and *CA$_{supernatant}$* indicates the identified crotonic acid in the supernatant by HPLC. As a consequence, a closer value to 0% indicates a better mass balance. In this study, most of the experiments had mass balance errors smaller than 5% (see Table 1).

A degree of PHA degradation was defined as the fraction of HB or CA concentration over total initial PHB mass in the biomass (equation 3 or 4).

$$HB/PHB_{initial} = \frac{HB_{supernatant}}{PHA_{initial}} \cdot 100\% \qquad [g/g]$$

(3)

or,

$$CA/PHB_{initial} = \frac{CA_{supernatant}}{PHA_{initial}} \cdot 100\% \qquad [g/g]$$

(4)

Chemical PHB degradation may occur either randomly in the middle of the polymer chain or from the end of the polymer chain. The GC method measured the overall lost PHB in the supernatant in terms of both soluble PHB oligomers and HB monomer, while HPLC method only quantified the HB monomers. A ratio between soluble HB monomer and overall PHB in the supernatant was used to assess the chemical PHB degradation mechanism (equation 5). A higher value (close to 1) indicates that HB is sole product of PHB degradation, suggesting PHB is degraded from the end of the polymer chain. Otherwise, PHB is more likely hydrolyzed by chemicals randomly from the middle of the chain, generating oligomers as products.

$$HB/PHB_{supernatant} = \frac{HB_{supernatant}}{PHA_{supernatant}} \cdot 100\% \qquad [g/g]$$

(5)

Fourier Transform Infrared Spectroscopy (FTIR)

The composition and the crystallinity of dry pellets were examined using a spectrum 100 FT-IR spectrometer (PerkinElmer). The solid powders were pressed on a germanium crystal window of a microhorizontal ATR for measurement of single reflection and absorption of infrared by the specimens.

Thermal Stability

Around 100 mg of an untreated biomass, PHB isolated from biomass by a chemical or an organic solvent treatment and/or a commercial PHB (Tianan, China) were isothermally treated in a compression molding machine (Dr Collins) at 170°C for a certain period of time (1, 3, 5, 10 and 15 min). The molecular weight of PHB before and after the thermal treatment was determined by a size exclusion chromatography (SEC). For SEC analysis, around 3 mg of a sample was dissolved in 1 ml hexafluoroisopropanol (HFIP) at room temperature overnight. The sample was subsequently filtered using 0.2 μm filter. Molar mass distribution was determined using a Waters model 510 pump and a Waters 712 WISP chromatograph with PL-gel mix D columns (300 × 7.5 mm, Polymer Laboratories). HFIP was used as an eluent with a flow rate of 1 ml/min. The system was calibrated with PMMA standards.

The thermal degradation rate can be expressed by the equation 6 (Grassie et al. [1984a], [b]):

$$\left(\frac{1}{P_{n,t}} - \frac{1}{P_{n,0}} \right) = k_D \, t \qquad [1/s]$$

(6)

Where, $P_{n,t}$ and $P_{n,0}$ are number average degrees of polymerization at time t and time 0 s, respectively. The rate constant (k_D) was determined from the slope of the equation 6 function. $P_{n,t}$ and $P_{n,0}$ were calculated using number average of molecular weight (M_n) in time t and time 0 s according to equations 7a and 7b.

$$P_{n,t} = \frac{M_{n,t}}{M_m} \qquad [(g/mol)/(g/mol)]$$

(7a)

$$P_{n,0} = \frac{M_{n,0}}{M_m} \qquad [(g/mol)/(g/mol)]$$

(7b)

M_m is the molecular weight of a PHB monomer unit, i.e. 86.09 g/mol.

RESULTS

PHB Recovery and Purification

Alkalis and surfactant were two chemicals used in this study in order to purify and recover PHB from fresh biomass. Initially sole NaOH treatments with different concentration and treatment time were conducted (see Table 1). The final product purity increased by increasing NaOH concentration or by the prolonged treatment time, but the recovery yield was negatively influenced by those two parameters. On the basis of the final product purity and recovery yield, the treatment with 0.2 M NaOH for 1 h was chosen as the standard condition (see Table 1). Under this standard condition, the final product purity and the recovery yield can reach 87% and 97%, respectively. In order to improve the purity from the standard condition, and to favor the sustainability of the process, different chemicals combinations were tested. With the purpose of improving the purity, surfactant was added to the standard condition to remove the cell materials further. With additional dosage of SDS to our standard condition, the purity can be improved up to 99% with a slight decrease in recovery yield (91%). NH_4OH was aimed at replacing NaOH, because it is potentially easier to be recycled than NaOH (van Hee et al. [2005]). However, significant decrease was observed in both purity (to 63%) and recovery yield (63%) when treating fresh biomass with 0.2 M NH_4OH.

Besides recovering PHB from fresh biomass, the effect of pre-treatment such as lyophilization or freezing, was also studied. These pre-treatments led to a higher purity in all cases and an improved recovery yield in sole SDS and NH_4OH treatment (see Table 1). For comparison purposes, recovery and purification by solvent extraction was also conducted in this study. Extraction with dichloromethane reached 98% purity from lyophilized biomass. However, the recovery yield was very low (55%) in this study.

Thermal stability of purified PHB

In order to utilize PHAs as thermoplastics, thermal stability is a crucial parameter. Thermoplastic polymers are usually processed at temperatures at least 10°C above their melting point and typical residential time in an extruder does not exceed one minute. The processing temperature of PHB is usually between 170 and 180°C. Therefore, the thermal stability of the samples was studied in terms of PHB degradation during the first minute at 170°C.

Number average of molecular weight (M_n) of PHB as a function of time during the thermal treatment is shown in Table 2. PHB isolated from biomass by a solvent method and the commercial PHB showed the highest thermal stability with less than 7% M_n drop within the first minute of the treatment ($\Delta M_{n,1}$). The resulting molecular weight after the processing was still acceptable for a plastic application ($M_n > 169$ kg/mol). The sample purified by 0.2 M NaOH or by 0.2 M NaOH and 0.2% SDS showed much more pronounced molecular weight decrease ($\Delta M_{n,1} > 70\%$). The consequent molecular weights were below 45 kg/mol. As compared to the chemically purified PHB, the degradation of the polymer in the untreated biomass was less detrimental ($\Delta M_{n,1} = 62\%$). The rate of the polymer chain scission, i.e. the degradation rate constant (k_D), was calculated from the slope of the kinetic function shown in Figure 1. Thermal stability results at 170°C are summarized in Table 3, in terms of a ratio between k_D of a specific sample and k_D of the commercial PHB reference ($k_{D,ref}$). It can be observed that both, the untreated biomass and the chemically purified PHB showed significant deterioration in terms of a faster degradation rate. On the other side, the solvent isolated PHB performed even better than the commercial sample.

Table 2: Molecular weight (number average Mn and weight average Mw)

and molecular weight change $\left(\left(\frac{M_{n,0} - M_{n,t}}{M_{n,0}}\right) \times 100\right)$ of various PHB samples as a function of thermal treatment at 170°C

Sample	Chemical treatment	PHB purity [wt.%]	Time of thermal treatment at 170°C [min]	Mn [kg/mol]	Mw [kg/mol]	$\left(\left(\frac{M_{n,0} - M_{n,t}}{M_{n,0}}\right) \times 100\right)$ [%]

Commercial PHB	-	99	0	182	647	0
			1	169	583	7
			3	175	541	4
			5	119	391	35
			10	150	435	18
			15	135	373	26
PHB from biomass	-	67	0	135	224	0
			1	51	111	62
			3	33	62	76
			5	30	42	78
			10	25	34	81
			15	19	25	86
	Solvent	99	0	915	1755	0
			1	883	1731	3
			3	824	1573	10
			5	771	1562	15
			10	516	1144	44
			15	560	1255	39
	0.2 M NaOH	85	0	119	315	0
			1	19	39	84
			3	8	13	93
			5	6	8	95
			10	2	3	98
			15	1.8	2.2	98
	0.2 M NaOH +0.2% SDS	95	0	163	484	0
			1	45	73	72
			3	14	23	91
			5	11	20	93
			10	4	8	98
			15	3	4	98

Water content in the samples was in between 0.01 and 0.02 wt.%.

Jiang et al.

Jiang et al. AMB Express 2015 5:5, doi:10.1186/s13568-015-0096-5

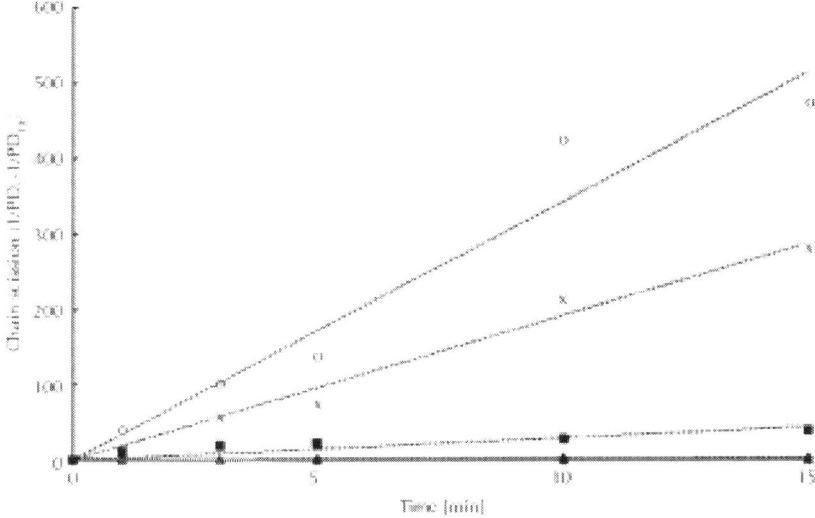

Figure 1: The effect of chemical treatment on thermal stability of commercial PHB (Tianan) and PHB isolated from biomass represented here by polymer chain scission ($1/P_{n,t} - 1/P_{n,0}$) as a function of time at 170°C. Water content in the samples was in between 0.01 and 0.02 wt%. The numbers in brackets represent PHB purity. PHB purified by 0.2 M NaOH (empty circle, 85% pure); PHB purified by 0.2 M NaOH and 0.2% SDS (cross, 95% pure); Unpurified PHB within biomass (solid square, 67% pure); PHB purified by solvent (empty triangle, 99% pure); Commercial PHB (solid diamond, 99% pure).

Table 3: Thermal degradation rate constants (k_D) of various PHB samples at 170°C and thermal degradation rate constants relative to the commercial PHB reference ($k_{D,ref}$) as a function of chemical treatment, purification method and purity

Sample	Chemical treatment	PHB purity [wt.%]	kD10−6[1/s]	kD/ kD,ref10−6[1/s]
Commercial PHB	-	99	0.18±0.02*	1.0
	0.2 M NaOH	99	1.40±0.10	8.0
	0.2% SDS	99	0.80±0.10	4.0

PHB from biomass	-	67	5.40±0.80	30.0
	Solvent	99	0.08±0.01	0.4
	0.2 M NaOH	85	54.00±5.00	300.0
	0.2 M NaOH+0.2% SDS	95	29.00±2.00	161.0

k_D of dried commercial PHB was used as a reference ($k_{D,ref}$).
Water content in all samples was in between 0.01 – 0.02 wt%.
Jiang et al.

Jiang et al. AMB Express 2015 5:5, doi:10.1186/s13568-015-0096-

PHB degradation by alkalis

The thermal instability of PHB purified by alkalis based method could be due to PHB hydrolysis. As it has been reported in Yu et al. ([2005]), abiotic hydrolysis of PHB by alkalis was observed in this study as well. Both HB monomer and CA were found as PHB hydrolysis products. Our data showed that the PHB degradation by NaOH in the fresh biomass was dependent on the treatment time and NaOH concentration. The hydrolysis products concentration showed linear relation with NaOH treatment time (Figure 2), while the relation between the NaOH concentration and the hydrolyzed products concentration is non-linear (Figure 3). In the tested NaOH concentration range, the HB monomer decreased with the increasing NaOH concentration before 0.1 M NaOH and then increased with NaOH concentration. For an initial PHB content of 68%, at the standard condition in this study (i.e. 0.2 M NaOH treatment for 1 h with fresh biomass), about 1.3% of initial PHB was hydrolyzed into HB monomer and about 0.6% of initial PHB was converted to CA.

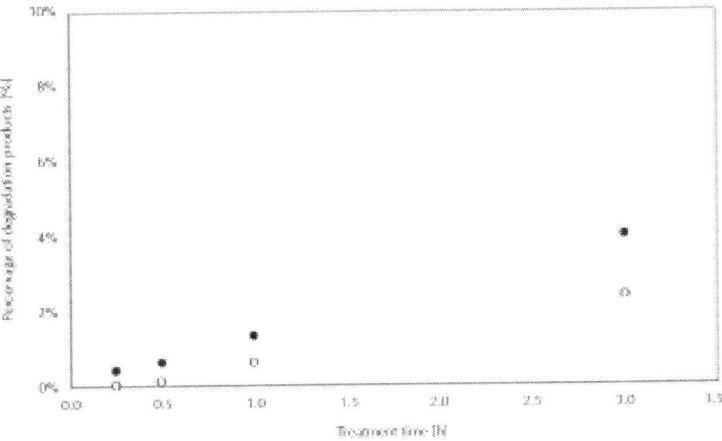

Figure 2: The relation between monomers production from PHB and NaOH treatment time. The fraction of two monomer products, hydroxybutyric acid (HB) and crotonic acid (CA) over total initial PHB (equations 3 and 4) are indicated by solid circle and empty diamond, respectively. The experiment was performed with fresh biomass at 0.2 M NaOH and 30°C in duplicate. Initial PHB content was 68%.

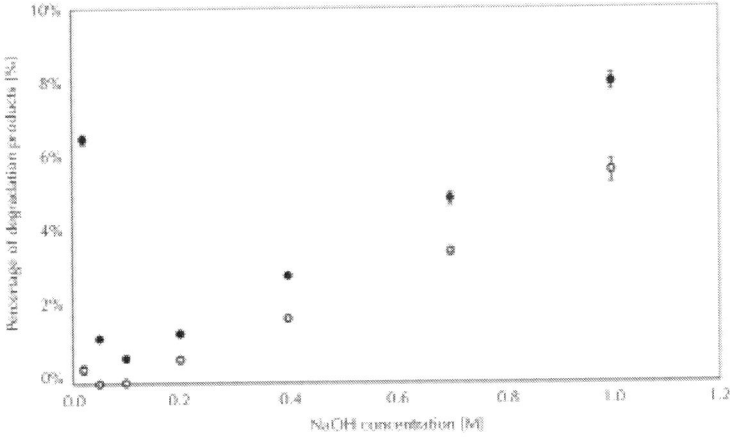

Figure 3: Fraction of degradation products, HB (solid round) and CA (empty diamond), over total initial PHB (equations3 and4).The experiment was performed with fresh biomass at 30°C for 1 hour in duplicate. Initial PHB content was 68%.

The pre-treatment step also showed some influence on the PHB hydrolysis. Much less HB or CA was produced after lyophilization or freezing pre-treatment (Figure 4).

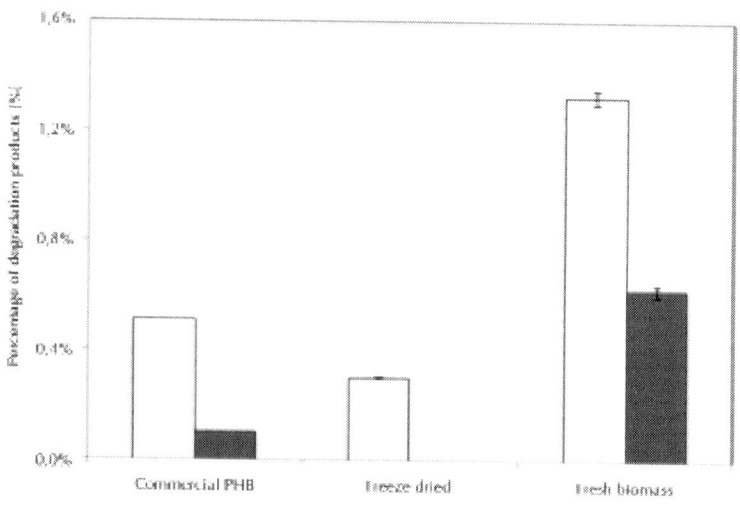

Figure 4: The influence of lyophilization on the PHB degradation by NaOH, expressed as fraction of degradation products over total initial PHB (equations3 and4). White color indicates HB and gray color represents CA. Samples were treated with 0.2 M NaOH for 1 hour at 30°C in duplicate. Initial PHB content was 68%.

The spectrum of hydrolysis products in the supernatant can be used as an indication of the chemical PHB degradation mechanism (equation 5). When the biomass with or without pre-treatment was treated by NaOH, $HB/PHB_{supernatant}$ ratio was always close to 100% (see Table 1). The closed mass balance in this study suggested that no other forms of soluble PHB oligomers were formed during NaOH treatment.

FTIR spectra

The effect of NaOH concentration and pre-treatment on PHB hydrolysis was investigated further by evaluating the crystallinity state of several samples through FTIR analysis (Xu et al. [2002]; Yu and Chen [2006]). An intensity ratio of the absorbance at 1230 cm^{-1} to that at 1453 cm^{-1} was used to calculate the polymer crystallinity index (CI, Xu et al. [2002]). Larger CI value corresponds to higher crystallinity whilst smaller values reflect lower crystalline portion. As can be seen from Table 4, both chemical treatment and pre-treatment process show influence on PHB CI value NH_4OH treated sample showed the lowest crystallinity compared to the rest of the samples.

Table 4: Crystallinity index $(CI = A_{1230}/A_{1453})$

Biomass status	Chemicals	CI
Commercial PHB	-	4.7
Lyophilized	CH2Cl2	5.7
Lyophilized	SDS	4.7
Lyophilized	NaOH	4.5
Fresh biomass	NaOH + SDS	4.4
Lyophilized	NH4OH	4.2
Fresh biomass	NaOH	3.8
Lyophilized	-	2.9
Fresh biomass	NH4OH	2.1

Larger value means that PHB is at a more crystallinity status and smaller value means that PHB is at a more amorphous status.

Jiang et al.

Jiang et al. AMB Express 2015 5:5, doi:10.1186/s13568-015-0096-5

FTIR can also be used to qualitatively detect both PHB and proteins in the final products (Yu and Chen [2006]). Therefore, all purified products were analyzed by FTIR, together with commercial PHB as control of PHB absorbance, and lyophilized biomass as a control of both PHB and protein absorbance. Figure 5a shows the spectrum of PHB from fresh biomass purified by different chemicals in comparison with commercial PHB and lyophilized biomass. The absorption at 1720 cm^{-1} and 1278 cm^{-1} respectively indicates C = O stretch and C-O stretch of the ester bonds. They both represent the presence of PHB. The absorption peaks at 1650 cm^{-1} and 1540 cm^{-1} represent amide I and amide II band in proteins. As can be seen in Figure 5, the commercial PHB and the PHB purified by NaOH-SDS mixture show highly similar spectra. In contrast, proteins were detected in all other samples.

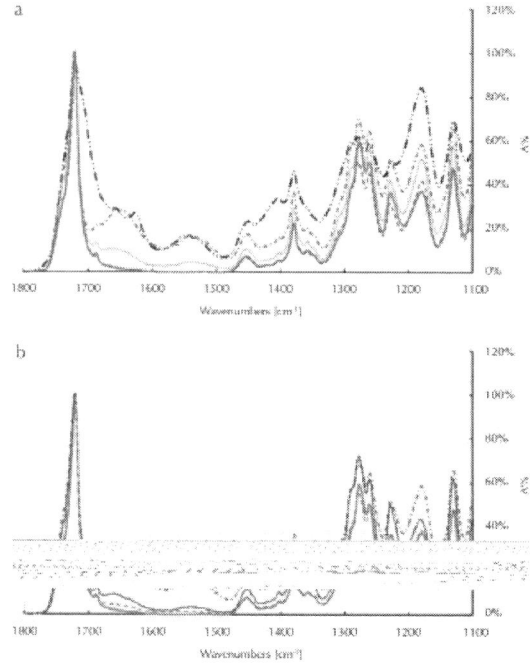

Figure 5: IR spectra of PHB-containing biomass treated with different chemicals. (a) Chemical treatment with fresh biomass. Freeze dried biomass, brown dash dot line; 0.2 M NH$_4$OH treatment, black dash double dots line; 0.2 M NaOH treatment, green dot line; 0.2 M NaOH and 0.25 w/v% SDS treatment, blue dash line; Commercial PHB, red line.(b) Chemical treatment with freeze dried biomass. Freeze dried biomass, brown dash dot line; 0.2 M NH$_4$OH treatment, black dot line; 0.2 M NaOH treatment, green dash line; Commercial PHB, red line. All of the treatments were conducted at 30°C for 1 hour. The absorbance intensity was normalized by the intensity of absorbance at 1720 cm^{-1}.

DISCUSSION

PHB Recovery and Purification

In this study, a high PHB purity was obtained from fresh biomass by treatment with alkali and surfactant. In principle, both alkali and surfactant can react with lipid and proteins, solubilizing the cell wall material and releasing the intracellular contents. Our results suggest that sole NaOH treatment can lyse cells but it is insufficient to remove all cell materials. Under our standard condition (0.2 M NaOH, for 1 hour at 30°C), still about 13.4% of cell material impurities remained in the final pellets. Those remaining impurities are likely water insoluble proteins and lipids. We observed that those hydrophobic impurities can be effectively removed by combined NaOH and SDS treatment. Higher SDS concentrations resulted in a higher final purity, likely due to micelle formation by SDS. Once the SDS concentration approached its critical micelle concentration (CMC), which is between 0.17-0.23 w/v%, more proteins and lipids were removed. However, SDS micelles might also solubilize PHB granules. Indeed, our data showed that PHB recovery yield decreased at high SDS concentration (Table 1).

The hydroxide ion concentration was also observed to have influence on cell materials removal and PHB recovery. In the case of NH_4OH and low NaOH concentration, for example, both purity and recovery were observed to be lower than at the standard condition. Since NH_4OH is a weak base, at the same solution concentration the amount of dissociated hydroxide ion from NH_4OH is much lower than for NaOH (about 100 times less). In fact, samples treated by 0.2 M NH_4OH and 0.02 M NaOH displayed the lowest purity in this study (respectively 63% and 77%, see Table 1). Next to a decreased removal of cell materials, treatment at 0.2 M NH_4OH showed more severe PHB degradation, which resulted in a lower recovery yield. This may be related to the PHB granules crystallinity status, which is discussed in the next section.

PHA recovery by chemical treatment has been widely reported in literature, but to our knowledge, on pure cultures only. The results are very diverse (Table 5). Considering the variability across studies in terms of microorganism, cell pre-treatment, temperature, initial PHB content and chemical concentration, among others, and their lack of PHB mass balance data, it is difficult to compare these results directly to our observations. Here we focus on the studies performed on fresh biomass, because at production scale it might be preferred to avoid any pre-treatment step.

Table 5: Literature comparison

Bacteria species	Biomass status	Chemical	Concentration	Initial PHA content	Purity	Recovery	Reference
E. coli(rec)	Frozen	NaOH	0.1	77%	91%	90%	Choi and Lee ([1999])
E. coli	Frozen	SDS	0.5%	77%	98%	87%	Choi and Lee ([1999])
E. coli	Frozen	NH4OH	0.1	77%	85%	95%	Choi and Lee ([1999])
E. coli	Frozen	H2SO4	0.1	77%	79%	87%	Choi and Lee ([1999])
C.necator	Lyophilized	NaOH	0.1	38%	97%	97%	(Mohammadi et al. [2012a],[b])
C.necator	Lyophilized	NaOH	0.1	60%	80%	90%	Anis et al. ([2012])
C.necator	Lyophilized	NH4OH	0.1	60%	60%	62%	Anis et al. ([2012])
Comamonas	Lyophilized	NaOH	0.05	34%	89%	97%	(Mohammadi et al. [2012a],[b])
R.eutropha	Lyophilized	NaOH	N.D.	70%	78%	45%	Yang et al. ([2011])

Species	State	Chemical	Conc.				Reference
R.eutropha	Lyophilized	SDS	5%	70%	90%	81%	Yang et al. ([2011])
R.eutropha	Lyophilized	SDS	1%	50%	87%	N.D.	Ramsay et al. ([1990])
P.putida	Lyophilized	NaOH	0.1	20%	40%	95%	Jiang et al. ([2006])
E. coli	Oven dried	NaOH+SDS	0.1+10%	60%	87%	96%	Peng et al. ([2013])
R.eutropha	Fresh	SDS	0.5%-20%	75%	97%	92%	Kim et al. ([2003])
R.eutropha	Fresh	H2SO4	1	60%	76%	94%	Yu and Chen ([2006])
A.vinelandii	Fresh	NH3	1	84%	94%	N.D.	Page and Cornish ([1993])
C.nector	Fresh	NaOH	0.1	68%	84%	91%	Anis et al. ([2013])
E. coli	Fresh	NaOH	0.2	79%	97%	91%	Choi and Lee ([1999])
P.acidivorans*	Fresh	NaOH	0.2	68%	89%	97%	This study
P.acidivorans*	Fresh	NH4OH	0.2	68%	65%	78%	This study
P.acidivorans*	Lyophilized	NH4OH	0.2	68%	87%	96%	This study
P.acidivorans*	Fresh	NaOH+SDS	0.2+0.2%	68%	99%	95%	This study

Dominant bacterial species in the mixed culture at the cultivation conditions of this study.

Jiang et al.Jiang et al. AMB Express 2015 5:5, doi:10.1186/s13568-015-0096-5

Choi and Lee ([1999]) reported that direct treatment of fresh recombinant *E.coli* by 0.2 M NaOH can result in 97% purity and 91% recovery. This is the best result described for sole NaOH treatment method. The major difference between their research and our study is that pure culture of recombinant bacteria were used in their research in contrast to mixed culture in our study. It is possible that some microorganism species in the mixed culture biomass are not efficiently treated by NaOH. Anis et al. ([2013]), for example, treated wet biomass of recombinant *C. necator* by 0.1 M NaOH, resulting in final purity (84%) and recovery yield (91%) more similar to our observations.

Regarding studies with sole surfactant treatment, Kim et al. ([2003]) applied SDS to *Ralstonia eutropha* cells, but additional heating at 121°C and washing steps were required to remove proteins and achieve a final purity of 97%. Interestingly, their PHB recovery (>92%) was remarkably higher than our results (63%, see Table 1). This suggests that temperature plays a significant role in the interaction between SDS and PHB – for example, due to altered critical micelle concentration (Bayrak [2003]) – resulting in less PHB loss with the supernatant.

The synergistic effect of alkalis and surfactants on PHB recovery and purification has not been well studied yet. Peng et al. ([2013]) combined SDS and NaOH for PHB purification of dried cells, resulting in lower purity (87%) but comparable recovery yield (96%) as in our study (99% and 95%, respectively).

PHB Degradation by Alkalis

We observed that a weak alkaline condition, 0.2 M NH_4OH and NaOH at concentration lower than 0.1 M, resulted in a larger degree of PHB hydrolysis. On the other hand, cell pre-treatment by lyophilization improved the recovery yield (Table 1) and resulted in less HB and CA monomers formed when compared to fresh cells (Figure 4). This effect may be related to the crystalline state of PHB granules during treatment. In the microbial cell, PHB granules are present as hydrophobic amorphous inclusions containing 5–10% of water (Yu and Chen [2006]). PHB granules at amorphous status are fragile to chemical hydrolysis. In fact, Yu and Chen ([2006]) and Valappil et al. ([2007]) suggested that PHB crystallization can increase PHB resistance to chemical treatment. PHB crystallization can be induced either by complete removal of water or by damaging the cell membrane (de Koning and Lemstra [1992]), the crystallization extent being dependent on the damage level of the membrane (Kawaguchi and Doi [1990]; Harrison et al. [1992]). Our results confirm their observations. At weak alkaline condition and without pre-treatment, PHB in the biomass seems to maintain its amorphous status (Table 4).

PHB hydrolysis decreases the molecular weight of final products, the rate and extent of decrease being dependent on the degradation mechanism. In this study, most of the lost PHB in the supernatant could be traced back in terms of HB monomer. Furthermore, a linear relation between HB concentration and treatment time also suggested that PHB degradation occurs at the end of the polymer chain (Figure 2). This is in agreement with the observations from Yu et al. ([2005]) on PHB from pure cultures.

Thermal Stability

Several studies have reported molecular weight and thermal properties as an indication of PHA quality for polymer applications, for PHAs obtained from pure cultures (e.g. Kim et al. [2003], Fiorese et al. [2009], Anis et al. [2012]) and from mixed cultures (summarized by Laycock et al.[2013]), and for several PHA recovery and purification methods. For thermoplastic applications, thermal stability is an important parameter. An instable polymer degrades during melt processing resulting in lower molecular weight material. At a certain critical molecular weight, mechanical properties substantially deteriorate. Kanesawa and Doi ([1990]) studied the effect of molecular weight on mechanical properties of PHBV copolymer. They reported that the tensile strength started to deteriorate at M_n of 50 kg/mol and at around 20 kg/mol the sample had no strength anymore. Hablot et al. ([2008]) studied the effect of fermentation residues, surfactants and processing conditions on both the thermal properties and thermal degradation of PHB obtained from pure cultures by a solvent method. To our knowledge, our study provides the first data on thermal stability of PHB obtained from mixed cultures.

The sample purified by solvent showed very similar thermal stability as compared to the commercial PHB, suggesting that the quality of PHB produced by the mixed microbial culture is comparable to PHB from pure cultures. On the other hand, PHB purified by chemical treatment showed severe thermal stability deterioration. By comparing the thermal degradation rate constants of several samples relative to the commercial PHB (Table 3), this effect could be attributed to 1) residues from the chemical treatment and 2) remaining biomass impurities. The inorganics used in the treatment could either attach to the polymer chain or stay as free molecules in the polymer. In both cases, they could catalyze a polymer chain scission either via β-elimination (Kim et al. [2006]) or hydrolysis mechanism (Yu and Marchessault [2000], Yu et al. [2005]). These results clearly show that the choice of recovery and purification method has a large impact on material properties.

In summary, this work studied the feasibility of purifying PHB from mixed culture biomass by alkaline-based chemical treatment. The purity and recovery obtained were comparable to those reported for pure cultures. PHB losses could be attributed to hydrolysis during the chemical treatment with HB monomer as main product, also in line with what has been observed for material from pure cultures. The extent of hydrolysis can be moderated by increasing the crystallinity of the PHB granules; in this study, by either adjusting the alkali concentration, or by cell pretreatment.

The recovery and purification method had a large influence on the quality of the product for thermoplastic applications. PHB purified by solvent displayed thermal stability comparable to commercial PHB. However, PHB obtained by alkaline treatment resulted in significant thermal stability deterioration, despite of the high purity and recovery yield obtained. The quality of the product for thermoplastic applications might be improved by further optimizing the alkaline treatment process, targeting residual inorganics and biomass components. Given the potential advantages of the alkaline treatment in terms of process economics and environmental impact, it is expected that this method can be of interest for other PHB applications.

AUTHORS' CONTRIBUTIONS

YJ, GM and MC participated in the design of the experiments and analysis of results. YJ performed the experiments. GM performed the thermal stability testing. YJ and GM wrote the manuscript. RK, LW, MC edited the manuscript. All authors read and approved the final manuscript.

ACKNOWLEDGEMENTS

These investigations were financially supported by the Technology Foundation STW (W2R, project nr. 11605). We thank Leonie Marang for kindly biomass supplies and Judith van Gorp and Lijing Xue for lab assistance.

REFERENCES

1. Albuquerque MGE, Torres CAV, Reis MAM (2010) Polyhydroxyalkanoate (PHA) production by a mixed microbial culture using sugar molasses: effect of the influent substrate concentration on culture selection. Water Res 44(11):3419-3433

2. Anis SNS, Iqbal NM, Kumar S, Amirul A-A (2013) Effect of different recovery strategies of P(3HB-co-3HHx) copolymer from *Cupriavidus necator* recombinant harboring the PHA synthase of*Chromobacterium sp* USM2. SepPurif Technol 102:111-117

3. Anis SNS, Nurhezreen MI, Sudesh K, Amirul AA (2012) Enhanced recovery and purification of P(3HB-co-3HHx) from recombinant *Cupriavidus necator* using alkaline digestion method. Appl Biochem Biotech 167(3):524-535

4. Bayrak Y (2003) Micelle formation in sodium dodecyl sulfate and dodecyltrimethylammonium bromide at different temperatures. Turk J Chem 27(4):487-492

5. Bengtsson S, Werker A, Christensson M, Welander T (2008) Production of polyhydroxyalkanoates by activated sludge treating a paper mill wastewater. Bioresource Technol 99(3):509-516

6. Bucci DZ, Tavares LBB, Sell I (2005) PHB packaging for the storage of food products. Polym Test 24(5):564-571

7. Chen G-Q (2009) A microbial polyhydroxyalkanoates (PHA) based bio- and materials industry. Chem Soc Rev 38(8):2434-2446

8. Choi JI, Lee SY (1997) Process analysis and economic evaluation for poly(3-hydroxybutyrate) production by fermentation. Bioprocess Eng 17(6):335-342

9.   Choi JI, Lee SY (1999) Efficient and economical recovery of poly(3-hydroxybutyrate) from recombinant *Escherichia coli* by simple digestion with chemicals. Biotechnol Bioeng 62(5):546-553

10.   de Koning GJM, Lemstra PJ (1992) The amorphous state of bacterial Poly[(R)-3-Hydroxyalkanoate] Invivo. Polymer 33(15):3292-3294

11.   de Koning GJM, Witholt B (1997) A process for the recovery of poly(hydroxyalkanoates) from Pseudomonads Part 1: solubilization. Bioprocess Eng 17(1):7-13

12.   Dias JML, Lemos PC, Serafim LS, Oliveira C, Eiroa M, Albuquerque MGE, Ramos AM, Oliveira R, Reis MAM (2006) Recent advances in polyhydroxyalkanoate production by mixed aerobic cultures: from the substrate to the final product. Macromol Biosci 6(11):885-906

13.   Dionisi D, Carucci G, Papini MP, Riccardi C, Majone M, Carrasco F (2005) Olive oil mill effluents as a feedstock for production of biodegradable polymers. Water Res 39(10):2076-2084

14.   Elbahloul Y, Steinbuechel A (2009) Large-scale production of Poly(3-Hydroxyoctanoic Acid) by *Pseudomonas putida* GPo1 and a simplified downstream process. Environ Microb 75(3):643-651

15.   Fiorese ML, Freitas F, Pais J, Ramos AM, de Aragao GMF, Reis MAM (2009) Recovery of polyhydroxybutyrate (PHB) from *Cupriavidus necator* biomass by solvent extraction with 1,2-propylene carbonate. Eng Life Sci 9(6):454-461

16.   Grassie N, Murray EJ, Holmes PA (1984) The thermal degradation of poly(−(D)-β-hydroxybutyric acid): Part 1—identification and quantitative analysis of products. Polym Degrad Stabil 6(1):47-61

17.   Grassie N, Murray EJ, Holmes PA (1984) The thermal degradation of poly(−(D)-β-hydroxybutyric acid): Part 2—changes in molecular weight. Polym Degrad Stabil 6(2):95-103

18.   Hablot E, Bordes P, Pollet E, Avérous L (2008) Thermal and thermo-mechanical degradation of poly(3-hydroxybutyrate)-based multiphase systems. Polym Degrad Stabil 93(2):413-421

19. Harrison STL, Chase HA, Amor SR, Bonthrone KM, Sanders JKM (1992) Plasticization of Poly(Hydroxybutyrate) Invivo. Int J Biol Macromol 14(1):50-56

20. Jacquel N, Lo C-W, Wei Y-H, Wu H-S, Wang SS (2008) Isolation and purification of bacterial poly (3-hydroxyalkanoates). Biochem Eng J 39(1):15-27

21. Jiang X, Ramsay JA, Ramsay BA (2006) Acetone extraction of mcl-PHA from *Pseudomonas putida*KT2440. J Microbiol Meth 67(2):212-219

22. Jiang Y, Marang L, Tamis J, van Loosdrecht MCM, Dijkman H, Kleerebezem R (2012) Waste to resource: converting paper mill wastewater to bioplastic. Water Res 46(17):5517-5530

23. Jiang Y, Sorokin DY, Kleerebezem R, Muyzer G, van Loosdrecht M (2011) *Plasticicumulans acidivorans gen. nov., sp nov.*, a polyhydroxyalkanoate-accumulating gammaproteobacterium from a sequencing-batch bioreactor. Int J Syst Evol Micr 61:2314-2319

24. Johnson K, Jiang Y, Kleerebezem R, Muyzer G, Van Loosdrecht MCM (2009) Enrichment of a mixed bacterial culture with a high polyhydroxyalkanoate storage capacity. Biomacromolecules 10(4):670-676

25. Kanesawa Y, Doi Y (1990) Hydrolytic degradation of microbial poly(3-hydroxybutyrate -co-3-hydroxyvalerate) fibers. Macromol Chem Rapid Commun 11(12):679-682

26. Kawaguchi Y, Doi Y (1990) Structure of native Poly(3-Hydroxybutyrate) granules characterized by X-ray-diffraction. FEMS Microbiol Lett 70(2):151-156

27. Kim KJ, Doi Y, Abe H (2006) Effects of residual metal compounds and chain-end structure on thermal degradation of poly(3-hydroxybutyric acid). Polym Degrad Stabil 91(4):769-777

28. Kim M, Cho KS, Ryu HW, Lee EG, Chang YK (2003) Recovery of poly(3-hydroxybutyrate) from high cell density culture of *Ralstonia eutropha* by direct addition of sodium dodecyl sulfate. Biotechnol Lett 25:55-59

29. Kunasundari B, Sudesh K (2011) Isolation and recovery of microbial polyhydroxyalkanoates. Express Polym Lett 5(7):620-634

30. Laycock B, Halley P, Pratt S, Werker A, Lant P (2013) The chemomechanical properties of microbial polyhydroxyalkanoates. Prog Polym Sci 38(3–4):536-583

31. (2013) Polyhydroxyalkanoate (PHA) Market, By Application (Packaging, Food Services, Bio-medical, Agriculture) & Raw Material — Global Trends & Forecasts to 2018. Report code:CH1610.

32. Mohammadi M, Hassan MA, Phang L-Y, Shirai Y, Man HC, Ariffin H, Amirul AA, Syairah SN (2012) Efficient Polyhydroxyalkanoate recovery from recombinant *Cupriavidus necator* by using low concentration of NaOH. Environ Eng Sci 29(8):783-789

33. Mohammadi M, Hassan MA, Shirai Y, Man HC, Ariffin H, Yee L-N, Mumtaz T, Chong M-L, Phang L-Y (2012) Separation and purification of Polyhydroxyalkanoates from newly isolated *Comamonas sp*EB172 by simple digestion with sodium hydroxide. Separ Sci Technol 47(3):534-541

34. Page WJ, Cornish A (1993) Growth of *Azotobacter-Vinelandii* Uwd in fish peptone medium and simplified extraction of poly-beta-hydroxybutyrate. Appl Environl Microb 59(12):4236-4244

35. Peng Y-C, Lo C-W, Wu H-S (2013) The isolation of poly(3-hydroxybutyrate) from recombinant*Escherichia coli* XL1-blue using the digestion method. Can J Chem Eng 91(1):77-83

36. Ramsay JA, Berger E, Ramsay BA, Chavarie C (1990) Recovery of poly-3-hydroxyalkanoic acid granules by a surfactant-hypochlorite treatment. Biotechnol Tech 4(4):221-226

37. Ramsay JA, Berger E, Voyer R, Chavarie C, Ramsay BA (1994) Extraction of poly-3-hydroxybutyrate using chlorinated solvents. Biotechnol Tech 8(8):589-594

38. Riedel SL, Brigham CJ, Budde CF, Bader J, Rha C, Stahl U, Sinskey AJ (2013) Recovery of poly(3-hydroxybutyrate-co-3-hydroxyhexanoate) from *Ralstonia eutropha* cultures with non-halogenated solvents. Biotechnol Bioeng 110(2):461-470

39. Serafim LS, Lemos PC, Torres C, Reis MAM, Ramos AM (2008) The influence of process parameters on the characteristics of polyhydroxyalkanoates produced by mixed cultures. Macromol Biosci 8(4):355-366

40. Sudesh K, Abe H, Doi Y (2000) Synthesis, structure and properties of polyhydroxyalkanoates: biological polyesters. Prog Polym Sci 25(10):1503-1555

41. Terada M, Marchessault RH (1999) Determination of solubility parameters for poly(3-hydroxyalkanoates). Int J Biol Marcromol 25:207-215

42. Valappil SP, Misra SK, Boccaccini AR, Keshavarz I, Bucke C, Roy I (2007) Large-scale production and efficient recovery of PHB with desirable material properties, from the newly characterised*Bacillus cereus* SPV. J Biotechnol 132(3):251-258

43. Van Hee P, van der Wielen LAM, van der Lans RGJM (2005) Method for the Production of a Fermentation Product from an Organism.

44. van Wegen RJ, Ling Y, Middelberg APJ (1998) Industrial production of polyhydroxyalkanoates using*Escherichia coli*: an economic analysis. Chem Eng Res Des 76(A3):417-426

45. Vishniac W, Santer M (1957) Thiobacilli. Bacteriol Rev 21(3):195-213

46. Williams SF, Martin DP (2002) Applications of PHAs in Medicine and Pharmacy. A Chapter in Biopolymers.

47. Xu J, Guo BH, Yang R, Wu Q, Chen GQ, Zhang ZM (2002) In situ FTIR study on melting and crystallization of polyhydroxyalkanoates. Polymer 43(25):6893-6899

48. Yang Y-H, Brigham C, Willis L, Rha C, Sinskey A (2011) Improved detergent-based recovery of polyhydroxyalkanoates (PHAs). Biotechnol Lett 33(5):937-942

49. Yu J, Chen LXL (2006) Cost-effective recovery and purification of polyhydroxyalkanoates by selective dissolution of cell mass. Biotechnol Progr 22(2):547-553

50. Yu J, Plackett D, Chen LXL (2005) Kinetics and mechanism of the monomeric products from abiotic hydrolysis of poly[(R)-3-hydroxybutyrate] under acidic and alkaline conditions. Polym Degrad Stabil 89(2):289-299

51. Yu G, Marchessault RH (2000) Characterization of low molecular weight poly(β-hydroxybutyrate)s from alkaline and acid hydrolysis. Polymer 41(3):1087-1098

Purification and Partial Characterization of a Thermostable Laccase from Pycnoporus Sanguineus CS-2 with Ability to Oxidize High Redox Potential Substrates and Recalcitrant Dyes

Sergio M. Salcedo Martínez[1, 2], Guadalupe Gutiérrez-Soto[1, 3], Carlos F. Rodríguez Garza[1], Tania J. Villarreal Galván[1], Juan F. Contreras Cordero[4], and Carlos E. Hernández Luna[1]

[1]Autonomous University of Nuevo León, Laboratory of Enzymology, Faculty of Biological Sciences, San Nicolás de los Garza, N.L. México

²Autonomous University of Nuevo León, Department of Botanic, Faculty of Biological Sciences, San Nicolás de los Garza, N.L. México

³Autonomous University of Nuevo León, Department of Biotechnology, Faculty of Agronomy, San Nicolás de los Garza, N.L. México

⁴Autonomous University of Nuevo León, Department of Microbiology and Immunology, Faculty of Biological Sciences, San Nicolás de los Garza, N.L. México

INTRODUCTION

Laccases (benzenediol: oxygen oxidoreductases, EC 1.10.3.2) are enzymes that catalyze the oxidation of phenolic compounds and aromatic amines with the simultaneous reduction of molecular oxygen to water [1]. They are widely distributed in many plants and fungi, some insects and bacteria, being particularly abundant in white-rot basidiomycetes [2]. Typical fungal laccases are described as glycosylated multicopper proteins, which are produced as extracellular monomeric forms of around 60-80 kDa, containing four copper atoms and 15-20% carbohydrates. Operatively, they are moderately thermotolerant, showing optima activity at 50-55 °C, and under acidic conditions (pH 3-5); although their maxima stability occurs in the alkaline zone (pH 8-9) [3]. Their copper atoms are distributed in three different sites bringing unique spectroscopic properties: The type 1 copper (CuT1) atom, is responsible of the intense blue color of enzymes by light absorption around 610 nm; The type 2 copper (CuT2) atom exhibits a weak absorption in the visible region; and the two type 3 copper (CuT3) atoms are present as a binuclear center, which has an absorption maximum about 330 nm. Moreover, CuT2 and CuT3 copper atoms are structural and functionally arranged as a trinuclear cluster. The four copper atoms form part of the active site of enzyme contributing directly to reaction. CuT1 is involved in the initial electron subtraction from reducer substrates, while trinuclear CuT2 and CuT3 cluster is responsible of the electron transference, from CuT1 to diatomic oxygen [4].

According to their redox potential, most of blue laccases belong to class II (-500 to -600 mV) or class III (-700 to -800 mV) laccases [5]. This is a disadvantage when compared to the ability of lignin peroxidase (LiP) and manganese peroxidase (MnP) to attack compounds with higher redox potential, including non-phenolic lignin units. To overcome this limitation laccases have evolutively developed a synergistic catalytic strategy, which combines a flexible ability to recognize a great variety of chemical compounds, with an extended capability to act at the distance through the activation of diffusible low molecular substances which serve as redox mediators. From a biological stand point this strategy let laccases to become one of the most versatile enzymes in nature, adaptable to multiple functions in plants, insects, fungi and bacteria. Another interesting possibility arises from the properties of atypical "yellow" and "white" laccases, which have shown the ability to catalyze the direct oxidation of high redox potential non-phenolic lignin model substrates or polyaromatic hydrocarbons [6, 7]. It has been proposed that the improved redox capabilities of these laccases come up either, by substituting some copper atoms for zinc, iron, or manganese in the metal clusters or by a change of the redox state of the CuT1(due to the interaction with a lignin-derived ligand) at the active site of, otherwise normal laccase protein structures. So, evolution and prevalence of laccases as a part of the lignin modifying enzyme (LME) system in white-rot basidiomycetes could also be the result of a "biochemical spring-up" mechanism acting under a short term ecophysiological selective pressure.

Whether directly or by mediation, laccases are able to oxidize a broad range of natural or xenobiotic compounds, including: mono, poly or methoxy- amine- and chloro-substituted phenols as well as aromatic heterocyclic and inorganic/organometallic substances; some of them recognized among the most recalcitrant industrial pollutants, for example; polycyclic aromatic hydrocarbons (PAH), pentachlorophenols (PCP), polychlorinated biphenyls (PCB), 1,1,1-trichloro-2,2-bis(4-chlorophenyl)ethane (DDT), trinitrotoluene (TNT), and many azo, triarylmethane, anthraquinonic, indigoid and heterocyclic textile dyes [8,9]. Therefore, laccases are considered enzymes with a great potential for the development of environmental and industrial applications. Current and potential laccase applications include biobleaching of pulp and bioremediation of pulp and paper industries, bioremediation of olive mill wastewater, bioremediation of effluents of the textile and dye manufacturing industries, biocatalytic synthesis of antibiotics and novel polymeric materials, development of biosensors, clarification and stabilization of beer, juices and wines, and panification [1, 10, 11]. Some laccase formulations have already reached a commercial significance, but general thought is that their biotechnological applications and performances could be greatly improved or expanded with the development and finding of new enzyme variants with desirable functional properties, such as higher redox potential, optimum activity at neutral or alkaline pH and thermal stability [12, 13, 14]. It has been proposed that these new laccases could be obtained by protein engineering or through the exploration of the natural biodiversity. The importance of prospective studies in natural biodiversity applying an ecophysiological approach is illustrated by reports about isolation of new thermostable laccases from fungi, either from thermophilic compost [15] or tropical environments [13,16], or by the finding of novel laccases with improved ability to oxidize substrates with a higher than normal redox potential culturing under solid phase conditions [6, 7,17]. Northeast Mexico shelters a high diversity of white-rot basidiomycetes as a result of its particular combination of physiography and climate, including species associated to pine, oak and mixed forests, sub-mountain and semi-desert scrublands, and grass-land. In this work we first present information on the isolation, identification and selection of a northeast Mexico native strain of *Pycnoporus sanguineus* CS2, as a potential producer of thermostable laccases. Results on the purification and partial characterization of its laccase are then exposed, stressing on its thermal stability and ability to attack high redox potential substrates and recalcitrant dyes without the participation of redox mediators.

MATERIALS AND METHODS

Chemicals

All chemicals used as buffers, enzyme substrates, culture media ingredients and electrophoresis reagents, were reactive grade and commercially available through local distributors of Difco, Sigma-Aldrich and Fluka, or BioRad products: PDB (potato dextrose broth), bacteriological agar, yeast extract, malt extract, peptone and dextrose, were from Difco. Acrylamide, bis-acrylamide, TEMED (N,N,N',N'-tetramethylethylenediamine), 2-mercaptoethanol, SDS (sodium dodecyl sulfate), trizma-base, glycine, Coomassie blue, and low range markers kit, were from Bio Rad. Enzyme substrates and dyes: 2,6-dimethoxyphenol (2,6-DMP); o-dianisidine (3,3'-dimethoxybenzidine); ABTS (2,2'–azino-bis(3-ethylbenzthiazolin-6-sulphonic acid); syringaldazine (4-hydroxy-3,5-dimethoxybenzaldehyde azine); DMAB (3-dimethylaminobenzoic acid), MBTH (3-methyl-2-benzothiazolinone hydrazone); Methyl Red (Acid Red 2; CI 13020); Reactive Black 5 (RB 5; CI 20505), were from Sigma, Fluka or Aldrich. Chromatographic matrices; DEAE-Sepharose and Q-Sepharose from Sigma-Aldrich, and Biogel P-100 from BioRad. All other chemicals, including solvents, inorganic salts, acids and bases, were from Reactivos Químicos Monterrey, S.A. or CTR-Scientific S.A. de C.V. Solutions and culture media were prepared with bidistilled water from Laboratorios Monterrey, S. A.

Isolation and Identification of Fungal Strain

The *Pycnoporus* strain used in our experiments was isolated from fruit bodies developing on decayed logs that were gathered in a man-disturbed sub-mountain scrubland around Monterrey, N.L. (Northeast México). Mycelia cultures were obtained by standard mycological techniques, according to the procedure previously described [18]. Briefly, small flesh sections were aseptically removed from inside the carpophores, and transferred to YMGA (0.4% Yeast extract, 1.0% Malt extract, 0.4% Glucose, 1.5% agar) plates, supplemented with 10% Tartaric Acid and 0.004% Benomyl. Plates were incubated at 28 °C, and those with extensive mycelia growth were analyzed under the microscope to confirm a successful isolation. Stock cultures were maintained by periodic transfers every two or three months on YMGA plates and kept refrigerated at 4 °C. Carpophore morphologic features, measurements and photographs were registered previous to dissections for microscopic examination, and identification was done by following the taxonomical keys in reference [19], and in [20] for genera of polypores and the most common macromycetes from Mexico, respectively.

Enzyme and Protein Assays

Laccase activity was determined by triplicate at 25 °C in 3 ml cuvettes, monitoring the increase in absorbance at A_{468} ($\varepsilon=49,600$ $M^{-1}cm^{-1}$), using a Shimadzu UV-VIS mini 1240 spectrophotometer and 2,6-DMP as substrate. The assay mixture contained 0.01ml enzymatic extract, 0.1 ml of 60 mM 2.6-DMP in 2.89 ml of 200 mM citrate-phosphate buffer at pH 4.0 [21]. One unit of laccase activity was defined as the amount of enzyme required to oxidize 1 μmol of 2, 6-DMP per minute at 25 °C. In some cases, it was necessary to assess the presence of lignin peroxidase (LiP) and manganese peroxidase (MnP). LiP activity was estimated by the H_2O_2-dependant veratryl alcohol oxidation to veratraldehyde as in reference [22] MnP activity was measured by the formation of Mn^{3+}-tartrate complex during the oxidation of $MnSO_4$ in tartrate buffer as in [23]. The protein concentration was estimated by the Bradford assay (Protein Assay Bradford of BioRad) with bovine serum albumin as standard.

Strain Selection and Enzyme Production

Isolated *Pycnoporus* strain was selected as a potential source of thermostable laccases in a preliminary screening with crude enzyme preparations. 250 ml Erlenmeyer flask, containing 50 ml of natural LME inducers containing Bran Flakes (BF) media (2% Bran Flakes® in 60 mM potassium phosphate pH 6.0) [24], were inoculated with three 0.5 cm diameter cylinders of mycelia taken from the border of a YMGA growing colony and incubated at 28 °C under agitation at 150 rpm. Aliquots (200 μl) were removed from cultures and the extracellular fluid was separated by centrifugation at 14 K (Eppendorf 5415 C). Enzyme activity in supernatants was determined with 2 mM 2, 6-DMP final concentration, as described above, after sample incubation at 60 °C during different times in a four hour period. This phase of study included four different *Pycnoporus* sp. (CS 2, CS 20, CS 43 and LE 90) strains from the native basidiomycete collection of our laboratory. Among them, *Pycnoporus sanguineus* CS 2 was selected on the basis of its ability to produce a thermostable 2,6-DMP oxidizing activity, and for showing apparently a single band of activity when incubated with 2,6-DMP and SGZ in a parallel native PAGE analysis of crude supernatants.

Enzyme production was evaluated in submerged liquid cultures on the natural containing laccase-inductors BF or a modified Kirk medium (MK) [25], with the following composition: 10 g l⁻¹ dextrose, 1.0 g l⁻¹ yeast extract, 5.0 g l⁻¹ peptone, 2.0 g l⁻¹ ammonium tartrate,

1.0 g l^{-1} KH$_2$PO$_4$, 0.5 g l^{-1} MgSO$_4$, 0.5 g l^{-1} KCl, and 1.0 ml of 100 X trace element solution (0.5 g EDTA, 0.2 g FeSO$_4$, 0.01 g ZnSO$_4$, 0.003 g MnCl$_2$, 0.03 g H$_3$BO$_4$, 0.02 g CoCl$_2$, 0.001 g CuCl$_2$, and 0.003 g NaMoO$_4$ in 100 ml); amended with 350 µM CuSO$_4$ and 3% ethanol [26]. Cultures were performed at 28 °C and 150 rpm for 14 days. 50 µL aliquots were taken every two days to determine the laccase activity.

Laccase Purification

All the procedures were performed at 4 °C, unless otherwise stated. Extracellular liquid from 14 day-old submerged cultures was separated from mycelium by filtration through a cotton-polyester 50:50% cloth. Then, water-soluble polysaccharides were removed from sample solution by freezing (- 20 °C for 24 h), thawing and filtration (Whatman # 1). Culture filtrate was concentrated to approximately 200 ml by 10 kDa ultrafiltration (Millipore prep/scale TFF cartridge). The obtained fluid was further reduced to 20 ml by using a stirred ultrafiltration system equipped with an YM10 membrane (Amicon, Millipore). The reddish-brown enzyme concentrate was equilibrated by diafiltration with 20 mM potassium phosphate, pH 6.0, and applied to a pre-equilibrated anion-exchange DEAE-Sepharose column (2.5 × 17 cm). Once on the column, unadsorbed protein and most of the pigment were removed by washing with two volumes of equilibrium buffer. Retained proteins were eluted with a linear gradient of potassium phosphate pH 6.0 from 20 to 300 mM, and the eluted fractions were assayed for laccase activity and the A$_{280}$ nm monitored. Fractions with laccase activity were pooled, concentrated, equilibrated by diafiltration with 100 mM potassium phosphate, and applied on a pre-equilibrated Biogel P-100 column (2.6 x 65 cm). The loaded proteins were eluted with the same buffer. Active fractions were pooled, concentrated and diafiltrated against 20 mM potassium phosphate buffer pH 6.0. Enzyme was further purified by anion-exchange on a pre-equilibrated Q-Sepharose column (2.5 x 17 cm). Once set the sample, active fractions were eluted with a lineal gradient of potassium phosphate from 20 to 300 mM. These fractions were pooled, concentrated and diafiltrated against water, and stored at - 20 °C.

Electrophoresis Analysis

Protein purity and molecular mass were evaluated by sodium dodecyl sulfate-polyacrylamide gel electrophoresis (SDS-PAGE) as in reference [27] with 4% stacking gel and 12% resolving gel. Protein bands were stained with Coomassie brilliant blue and the molecular mass (M_r) of purified laccase was determined by calculating the relative mobility of standard protein markers: *phosphorylase b* 97.4 kDa; *serum albumin* 66.2 kDa; *ovalbumin* 45 kDa; *carbonic anhydrase* 31 kDa; *lysozyme* 14.4 kDa; *aprotinin* 6.5 kDa (SDS-PAGE molecular weight standards low range, BioRad). Native PAGE was carried out as described at reference [28]. Activity staining of laccase was performed by incubating with 2, 6-DMP or the pair MBTH + DMAB in 200 mM sodium acetate buffer, pH 4.5. For identification and comparing proposes, the corresponding laccase band was removed from a parallel gel and submitted to the Proteomic Unit of IBT-UNAM at Cuernavaca, Morelos for aminoacid sequencing; resulting in the sequencing of six internal peptides.

UV-VIS Absorbance Spectra

As a part of the characterization of the physicochemical properties of the laccase, its absorbance spectrum from 200 to 800 nm was obtained in a UV-Vis Shimadzu-Mini 1240 spectrophotometer. The assay was performed with 25 µM of protein diluted in 1 ml of bidistilled water.

Effect of PH on Enzymatic Activity

Optimum pH of activity was determined in McIlvine buffer (consisting of a combination of 100 mM citrate/50 mM potassium phosphate) adjusted in a range from 3.0 to 7.0. Activity determination was made according to described method with 2,6-DMP using 0.2 M citrate phosphate buffer, at pH 4.0.

Effect of Temperature on Enzyme Activity and Stability

The effect of temperature on reaction rate was determined using 2, 6-DMP in 0.2 M citrate/phosphate buffer, at pH 4.0. The temperature of reaction mixture was adjusted to indicate value and then the reaction was started by the addition of enzyme. The assays were done by triplicate and data in graphics appear as relative activity as a function of temperature, considering as 100% the average of maxima obtained. The activation energy of the system was calculated by the Arrhenius model, according to the expression: Log k = [$-E_a$ / 2.303R (1/T)] + Log A, where: k is the rate constant (it depends of temperature); A is preexponential factor or frecuency factor. E_a the activation energy (expressed in J/mol); R is the gas universal constant (8.314 J K^{-1} mol^{-1}), and T the absolute temperature (°K). In the thermostability assays, the enzyme was pre-incubated at 50, 60 and 70 °C for the indicated periods of time and activity was measured at 25 °C on 2, 6-DMP in 0.2 M citrate/phosphate buffer, at pH 4. The assays were carried out by triplicate and data are expressed as percent of remaining activity as a function of incubation time; taking as 100% the average value of activity at time zero for each temperature. Inactivation process was adjusted to an exponential decay model, from which the constants of heath inactivation (k) and the half-life times were calculated according to the expression: ln (N_0/N) = kt, where; N_0 is the activity at the starting of incubation; N is remaining activity after a certain incubation time (t); k correspond to the first-order inactivation constant, and t$1/2$ is the half-life time, calculated as ln 2/k = 0.693/k

Determination of Kinetic Parameters

Kinetic analysis was performed on some common substrates of laccase: 2, 6-DMP, ABTS, o-dianisidine and SGZ. Reaction mixtures were prepared in 0.2 M citrate/phosphate buffer at pH 4.0 and the change in optical density by minute was measured by triplicate at different substrate concentrations (0.05, 0.1, 0.5, 1.0, 5.0, 10.0 mM) or those indicated for each assay. The assays were performed at 468 nm for 2, 6-DMP (ε = 49,600 M^{-1} cm^{-1}), 436 nm for ABTS (29,400 M^{-1}cm^{-1}), 460 nm for o-dianisidine (11,000 M^{-1} cm^{-1}) and 525 nm for SGZ (ε= 65 000 M^{-1} cm^{-1}). The values of the Michaelis constant (Km), maximum velocity ($Vmax$), turnover number ($Kcat$) and specificity constant $Kcat/Km$ were estimated according to the Lineweaver-Burk method.

Decolorization Assays

The decolorizing ability of laccase was evaluated with two recalcitrant dyes, the non-phenolic azo Methyl Red (MR), and the diazo reactive black 5 (RB 5). The reaction mixture consisted of 0.890 ml of 0.2 M citrate/phosphate buffer, pH 4.0, 0.1 ml of 250 µM MR or RB5 (final concentration 25 µM), and 0.01 ml of pure laccase (final concentration 5 U/ml). Assays were performed at 25 °C and reaction was initiated with the addition of enzyme. Decolorization was estimated by the decreasing of absorbance at 530 nm for MR or 597 nm for RB5. The results are expressed as the percent of remaining color as a function of incubation time according to the relationship: remaining color (%) = [(*Abs final/ Abs initial*)]*100, where: *Abs final* correspond to the absorbance value at the indicated incubation times, and *Abs initial*: is the initial (t = 0) absorbance value.

RESULTS AND DISCUSSION

Strain Identification

In this study an autochthonous strain of *Pycnoporus* sp (CS 2) was selected as a potential source of thermostable laccases for its ability to produce a thermotolerant 2, 6-DMP oxidizing activity in preliminary assays with crude filtrates from submerged cultures. This basidiomycete was initially isolated from fruit bodies, growing on decayed logs in a man disturbed sub-mountain scrubland around Monterrey, N.L. México (Figure 1), and identified by its morphological and microscopic traits. According to their morphological and microscopic characteristics, the carpophores corresponded to the species *Pycnoporus sanguineus* (L.) Murrill, for their bright orange to orange-red, red or cinnabar-red shelf-like basidiomes, which are nearly round to elongated or fan-shaped in outline, have a dry surface, smooth or finely hairy, wrinkled or warty and attain 2-12 cm diameter and 0.2 to 0.5 cm thick. Their margins are thin and the under surfaces are covered by small pores (3-4 per mm), bright orange to orange red or red ranging from 0.5-1.5 mm long. Their white spores are smooth and oblong-elliptical in shape and range from 4.2 to 5.2 microns long by 2 to 3.5 microns width and the flesh is tough, red to yellowish red, staining black with KOH. On the bases of these features, we assigned the strain under study as *Pycnoporus sanguineus* CS 2.

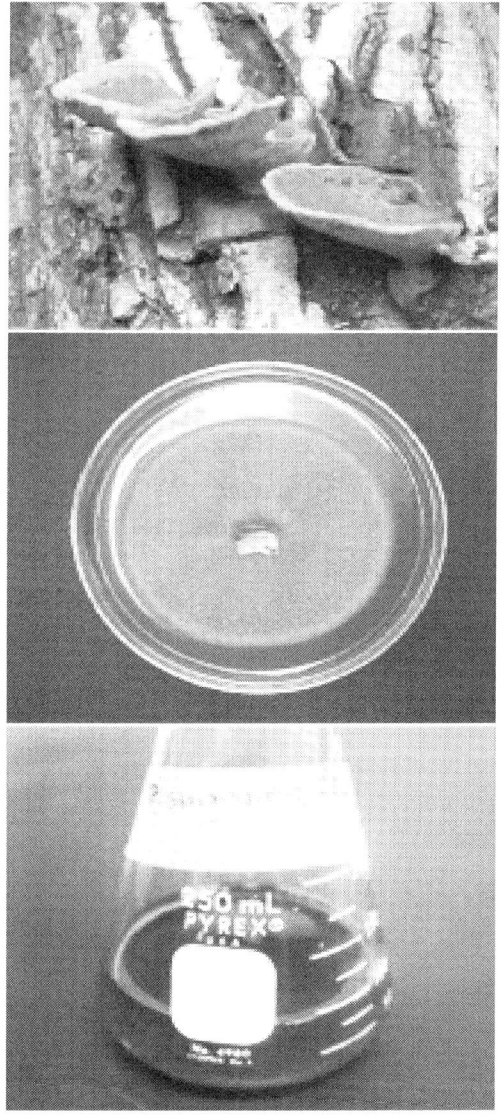

Figure 1: Fruit bodies (carpophores), mycelium colony and submerged culture of *Pycnoporus sanguineus* CS 2. Fungus identification was performed ac-

cording to macroscopic and microscopic features. Strain isolation was done by tissue transference from the inner flesh of carpophores using mycological standard methodologies. Develop of orange-red pigmented mycelium on the edge of the solid plate colonies, and extracellular production of a reddish pigment under submerged conditions, were indicative of a successful isolation. Production of extracellular mucilage was also observed on submerged cultures. Isolation and identification details are given in text.

Production and Purification of Laccase

Guzmán (2003), considers that *P. sanguineus* is a tropical variant of the temperate zone species *P. cinnabarinus*, adapted to man disturbed sites, where it is common in fallen logs and fences, always in sunny places [29]. As its closely related species, *P. cinnabarinus* and *P. coccineus, P. sanguineus* is recognized as an efficient lignin decomposer, in spite of its relatively simple LME system composed of laccases [5]. These features make *Pycnoporus* species an attractive group of white-rot basidiomycetes for the production and purification of unusual laccases [16, 30]. In this study, laccase production was carried out in submerged liquid cultures on a modified Kirk basal medium (MK), amended with 3.5 mM $CuSO_4$ and 3 % ethanol, as chemical laccase inducers and on Bran-Flakes medium (BF), containing natural LME inducers. Under these conditions maxima volumetric productions were reached in both media after 14-16 days (Figure 2). As laccase titers on MK media were about thrice higher than that on BF media (7.5 U ml^{-1} vs 2.3 U min^{-1}), it was selected for enzyme production in purification assays. Consistently with other reports on LME production by *Pycnoporus* species, LiP and MnP were not detected [5, 16, 30, 31, 32].

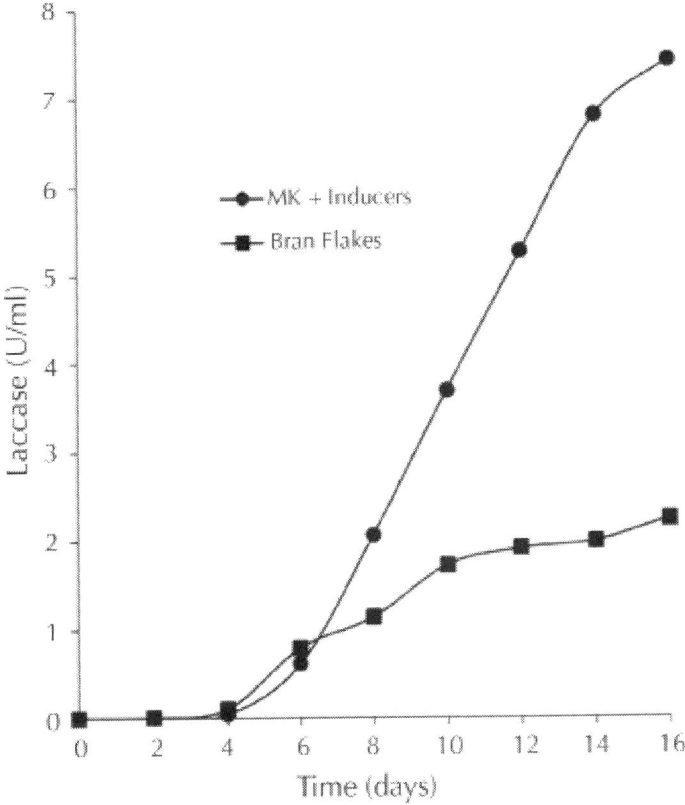

Figure 2: Time course of laccase production by *Pycnoporus sanguineus* CS2. Cultures were carried out on a modified Kirk basal medium, amended with 3.5 mM $CuSO_4$ (MK+ Inducers) and on Bran Flakes medium (BF) at 28 °C under 150 rpm agitation. Data represent the average of a representative assay in triplicate. Ethanol was added to MK medium at the third day of culture. Activity was determined with 2 mM 2, 6-DMP in 200 mM citrate-phosphate buffer pH 4.0.

Laccase purification was started from about 1850 ml of mycelium-free filtrates from 14 day-old submerged cultures. After 10K ultraconcentration and sequential steps of anionic exchange chromatography on DEAE- Sepharose, gel filtration on Biogel P-100, and anionic exchange on Q-Sepahrose, laccase activity eluted as an apparently single protein peak with 100-140 mM phosphate (Figure 3). When aliquots of pooled laccase from this last chromatographic step were analyzed by denaturing SDS-PAGE, multiple protein bands were detected by Coomassie staining (not shown). A similar effect was reported in a work with a *Fusarium proliferum* laccase. As the multiband effect persisted after SDS substitution by other detergents, but disappeared in the absence of SDS, this phenomenon was associated to the presence of detergent on denaturing PAGE [33]. However, when we applied both a heat denatured sample and a non-boiled sample in parallel using the same SDS gel, a protein multiband and a single band were detected by Coomassie staining, respectively. Furthermore, the simple band pattern was also obtained in duplicates by activity staining using laccase substrates. This indicates that thermal treatment could be responsible of the observed multiband effect on denaturing conditions, instead of the SDS by itself. A summary of purification data is shown in Table 1. By this procedure a 16.7-fold purification and activity recovery of 25.5%, with specific activity of 69 U mg^{-1} protein was achieved. Concentrated purified enzyme showed the blue color characteristic of multi-cupper oxidases.

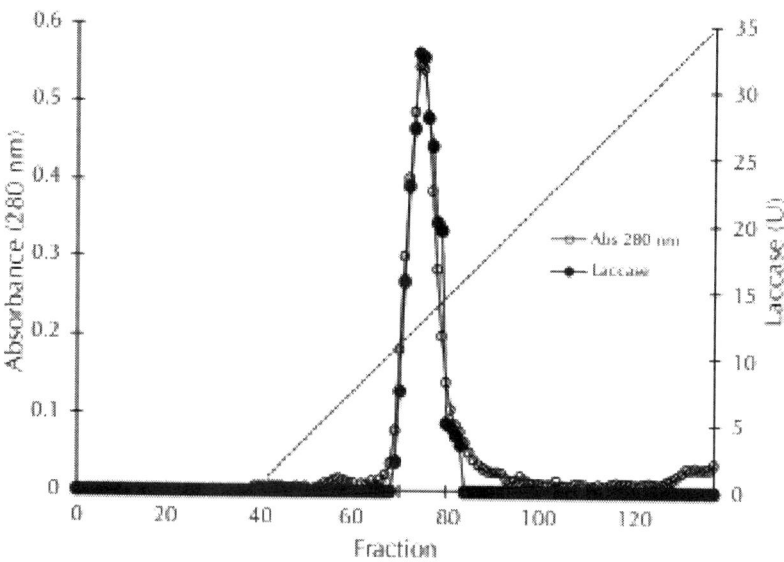

Figure 3: Elution profile for laccase from *Pycnoporus sanguineus* CS2 on anion-exchange column chromatography with Q-Sepharose (2.5 x 17 cm). The enzyme was eluted with a potassium phosphate (pH 6.0) linear gradient from 20-300 mM (dashed line) at a flow rate of 1.0 ml/min.

Table 1: Purification of *Pycnoporus sanguineus* CS 2 laccase

Purification step	Protein (mg)	Enzyme Activity (IU)	Specific Activity (U/mg)	Recovery (%)	Purification (fold)
Culture filtrate	1551	6477	4.18	100.0	1.0
Ultraconcentration 10 K	422	3724	8.8	57.4	2.1
DEAE-Sepharose FF	64	2308	35.8	35.6	8.5
Biogel P-100	32	1924	59.2	29.7	14.1
Q-Sepharose	22	1522	69.8	23.5	16.7

Biochemical Properties

Electrophoresis analysis indicated that *Pycnoporus sanguineus* CS 2 produced only one laccase under the conditions used in this study. According to non-denaturing SDS-PAGE this laccase is a monomeric protein with a molecular weight of 64.4 kDa, and activity staining of native gels exhibited a single broad band when incubated with both, 2,6-DMP and the pair MBTH + DMAB, showing the same migration as the Coomassie blue stained band (Figure 4). Molecular mass of purified laccase was very similar to those reported for different *Pycnoporus sanguineus* strains [30, 31, 32], and it was consistent with the reported for most of basidiomycetes laccases [2, 3]. As expected for its visual appearance described above, the UV-Vis spectrum of purified enzyme was characteristic of the typical blue laccases, displaying the absorbance peak near to 600 nm related to the Cu-T1 centers, and the shoulder at 330 nm of Cu-T3 binuclear centers (Figure 5). Nonetheless, the oxidative coupling of MBTH and DMAB in the absence of mediators was indicative that *Pycnoporus sanguineus* CS2 laccase has the capability to catalyze reactions requiring a higher than usual redox potential for typical laccases [15].

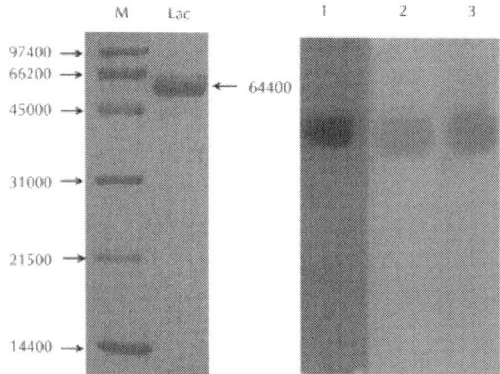

Figure 4: Electrophoresis analyses of purified *Pycnoporus sanguineus* CS 2 laccase by non-denaturing SDS- PAGE (left panel) and Native PAGE (right panel). Lanes M and Lac correspond to the Coomassie staining of molecular weight markers and purified laccase, respectively. The markers were phosphorylase b (97.4 kDa), serum albumin (66.2 kDa), ovalbumin (45 kDa), carbonic anhydrase (31 kDa), trypsin inhibitor (21.5 kDa), and lysozyme (14.4). On the right, lane 1 shows the Coomassie staining of purified laccase, and lanes 2 and 3, the activity staining with 2,6-DMP and the pair MBTH + DMAB, respectively.

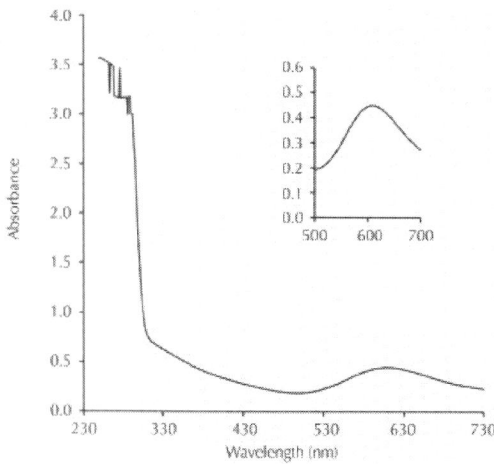

Figure 5: UV-Vis Spectrum of *Pycnoporus sanguineus* CS 2 Laccase. Assay was performed with a preparation of 25 µM laccase in bidistilled water. Insert shows the enlargement of the peak close to 610 nm.

In addition to blue laccases, other "atypical" forms of the enzyme named "yellow" laccases and "white" laccases have been reported. In the first case, it has been proposed that a variation in the redox state of Cu-T1 centers, by the presence of endogenous ligands, decreases the absorbance at 600 nm, without altering the spectral characteristics of the Cu-T2 and Cu-T3 centers, resulting in a yellow color [6]. In white laccases, like the one produced by *Pleurotus ostreatus*, it has been informed the presence of a single copper atom, which is accompanied by two of zinc and one of iron, instead of the regular four copper atoms [7]. In both, yellow and white laccases, the protein structure is similar to that found in blue laccases, but the changes in the redox state of the active site (whereas by the presence of the endogenous mediator or by the substitutions in the Cu centers), enables them to oxidize directly substrates of higher redox potential. According to all the above, the laccase produced by *P. sanguineus* CS 2 corresponds to a blue laccase, most likely containing the regular composition of Cu in its catalytic centers, but like atypical laccases is capable of acting on substrates of higher redox potential.

PH and Temperature Dependence

Enzyme was further characterized for its pH and temperature dependence. The effect of pH on laccase activity was studied using some of the most common laccase substrates, including the phenolic 2, 6-DMP, *o*-dianisidine and SGZ, as well as the non-phenolic ABTS. In general, laccase exhibited optima activity in the zone of pH between 3.0 and 4.5, depending on the particular substrate, then it declined in a gradual way towards the neutral zone of pH, and was completely lost at pH 6.5. Optimal pH values were 3.5, 3.5, 4.5 and 3.0 for 2, 6-DMP, *o*-dianisidine, SGZ and ABTS, respectively (Figure 6). These results were similar to those reported in literature for most of the fungal laccases [2]. It is known that biphasic pH-activity profiles with phenolic substrates (as the one showed by SGZ), are a consequence of two opposite effects: one generated by the difference in the redox potential between the reducer substrate and the Cu-T1 centers, when changing from acidic to neutral conditions. The other one is directly associated to the inhibitory action of OH⁻ ions over the activity of Cu-T2/T3 centers. For non-phenolic substrates as ABTS, the first effect should be minimal and the inhibition by OH⁻reflects the monotonic decrease in the enzyme activity [12, 34].

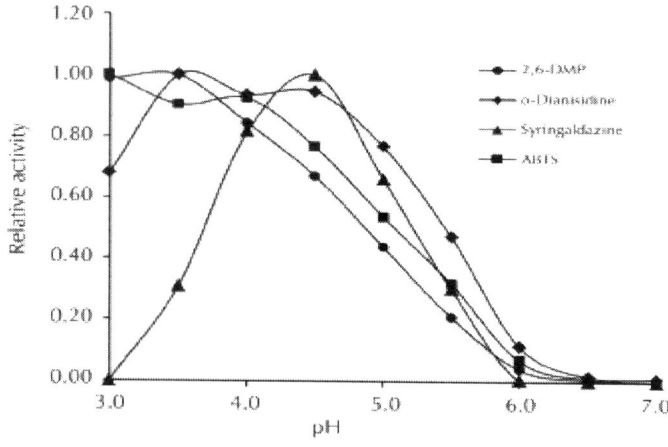

Figure 6: pH versus activity profiles of *Pycnoporus sanguineus* CS 2 laccase on various substrates. Assays were done by triplicate in 200 mM citrate-phosphate buffer at indicated pH, with 2, 6-DMP (2 mM),*o*-dianisidine (0.66 mM), SGZ (0.05 mM) and ABTS (2 mM).

The influence of temperature on *P. sanguineus* CS 2 laccase activity was investigated with 2, 6-DMP (2 mM), at pH 4.0, in the zone, from 20–80 °C. The profile temperature-activity showed a gradual increase from the lower limit at 20 °C to achieve an optimal value at 65 °C, and declined as temperature approached 80 °C. However, the enzymatic activity in these conditions remained relatively high compared to the value showed under optimal conditions (with a level close to 85%) (Figure 7). Indicating that *P. sanguineus* CS 2 laccase is a thermotolerant enzyme [35]. These data were evaluated according to the Arrhenius model in order to estimate the energy of activation (E_a) for the system. This parameter has been relatively little studied in thermotolerant laccases. The calculated E_a value (16.2 kJ/mol) for *P. sanguineus* CS 2 laccase is similar to the values reported for other thermotolerant laccases, as that for *Myceliophora thermophila* (19 kJ/mol) [36] and for the recombinant laccase from*Coprinus cinereus* (14 kJ/mol) [37], but smaller than those calculated in this report for other laccases, which apparently did not show a direct relationship between thermotolerance and the magnitude of Ea. On the other hand, the function showed a change in slope in the high temperature zone (50-70 °C) before the enzyme denaturing breaking zone. This effect could correspond to a decrease in the E_a of the system, caused by a thermotropic transition of the protein conformation, which should facilitate the limiting step of the reaction. Other possibility would be the coexistence of two enzyme populations, one of them showing an increased activity by temperature and the other being totally inactivated by thermal denaturing. These alternatives should be further explored.

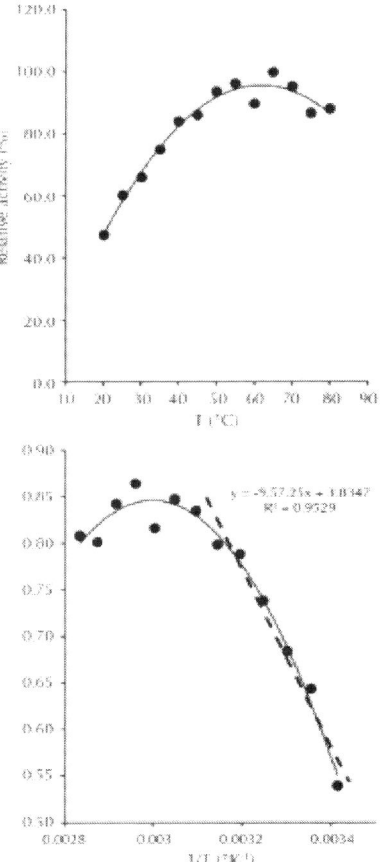

Figure 7: Effect of temperature on *Pycnoporus sanguineus* CS 2 laccase activity. Assays were performed by triplicate in 200 mM citrate-phosphate buffer at pH 4.0. Reaction rates were measured under saturating substrate concentrations (2 mM 2,6-DMP). The fitting line in lower panel shows the results of

the Arrhenius analysis of data, corresponding to: LOG (Vmax) = $[(-E_a/2.303\ RT) + constant]$.

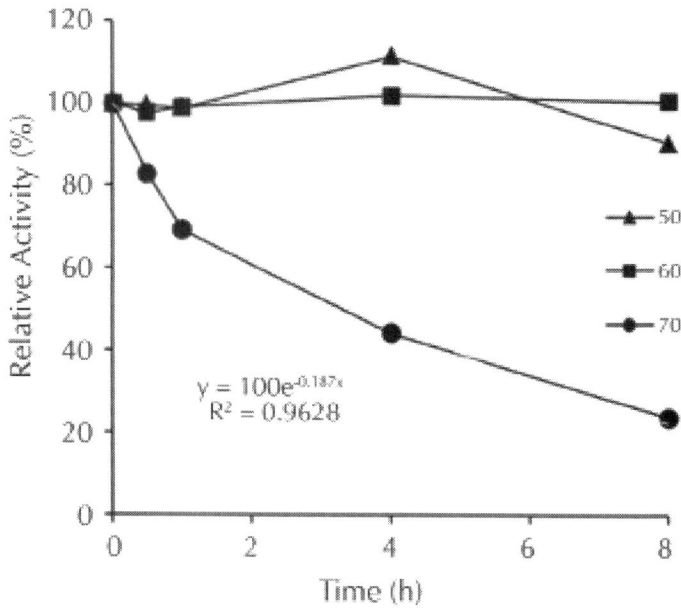

Figure 8: Effect of temperature on *Pycnoporus sanguineus* CS 2 laccase stability. Incubations were performed at various temperatures in distilled water. Aliquots were withdrawn at the indicated times and initial rates measured at 25 °C in 0.2 mM citrate/phosphate buffer at pH 4.0 with 2 mM 2, 6-DMP.

Thermostability is a desirable property of industrial enzymes. Curves of temperature-stability of *P. sanguineus* CS 2 laccase showed that enzyme retained practically all of its activity after incubation for 8 h at 50 and 60 °C. Moreover, when incubations at 60 °C were extended to 24 h, the laccase retained 98% of its original activity (not shown). The enzyme also retained almost 50% of its activity after 4h at 70 °C. Inactivation curve showed a first-order decaying behavior (correlation > 0.96), with a calculated half-life $(t_{1/2})$ of 3.85 h [corresponding to a constant of thermal inactivation (k) of 0.187 h^{-1}]. To the best of our knowledge, this is one of the highest $t_{1/2}$ values found in laccases from mesophilic fungi. It is known that most of typical fungal laccases lose their activity in a few minutes at 60 °C [3, 15, 30, 38].

In comparison to laccases isolated from other *Pycnoporus* species, $t_{1/2}$ value at 70 °C here described is higher than those reported for laccase I (0.13 h) and laccase II (2.06 h) from *Pycnoporus* sp SYBC-L1 [13, 30], and for the laccase from the thermotolerant *P. sanguineus* CelBMD001 (0.21 h) [16]. Native laccase also seems to be more resistant to thermal inactivation than *P. sanguineus* SCC 108 laccase ($t_{1/2}$= 3.33 h at 65 °C) reported by [31] and the *P. sanguineus* CCT-4518 laccase studied in [39], which lost 60% of its initial activity after 2 h at 70 °C. Interestingly, three laccases from tropical or subtropical strains of *Pycnoporus* species (*P. sanguineus* BRFM 902, *P. sanguineus* BRFM 66, and *P. coccineus* BRFM 938) of different geographic regions (French Guinea, China and Australia, respectively) with remarkable thermal resistance have been recently reported by a research group in France [13]. A relationship between *P. sanguineus* CS 2 laccase with these and other *Pycnoporus* laccases already described was established by comparing the aminoacid sequences of an internal protein fragment (peptides 2+3+4, Table 2) from the native laccase with those sequences deposited at GenBank. Aminoacid sequence of *P. sanguineus* CS 2 laccase showed 99 % similarity to *P. sanguineus* BRFM 902 laccase, 93% to *P. coccineus* BRFM 938 [13], *P. cinnabarinus* PM laccases [5], and *P. sanguineus* BRFM 66 laccase [13], but only 84 % to *Trametes cinnabarina* [40] and 71% to *P sanguineus* CelBMD001 laccase [16]. These results highlight the importance of *Pycnoporus* species biodiversity for the prospection for new thermostable laccases.

Kinetic Properties

The kinetic properties of enzyme were studied with some typical substrates. The values of the Michaelis constant (K_m), catalytic constant (K_{cat}) and specificity constant (K_{cat}/K_m), were calculated by the Lineweaver-Burk method. Laccase showed the highest affinity and molecular activity, on ABTS (K_m = 23 mM, K_{cat} = 221 s^{-1}) compared to o-dianisidine (K_m = 44 mM, K_{cat} = 197 s^{-1}), and 2, 6-DMP (K_m =41 mM, K_{cat} = 88 s^{-1}). So, in terms of catalytic efficiency the best substrate resulted ABTS (K_{cat}/K_m = 9.4 x 10^6 s^{-1} M^{-1}) followed by o-dianisidine (K_{cat}/K_m = 4.5 x 10^6 s^{-1} M^{-1}) and 2, 6-DMP (K_{cat}/K_m = 2.2 x 10^6 s^{-1} M^{-1}). These results are summarized in Table 3, comparing the values of specificity constants (*Kcat/Km*) for these substrates with those reported for other

*Pycnoporus*laccases, the native enzyme showed higher values for all assayed substrates except for the reported laccase II from *Pycnoporus sp* SYBC-L1 [30]. Like typical laccases, the enzyme of *P. sanguineus* CS 2, showed activity on a variety of substrates, such as the phenolic 2, 6-DMP, *o*-dianisidine and SGZ, as well as the non-phenolic ABTS.

Table 2: Amino acid sequences corresponding to internal peptides of *Pycnoporus sanguineus* CS2 laccase

Peptide	Amino acid sequences
1	EAVVVNGITPAPLIAGKK
2*	GPFVVYDPNDPQASLYDIDNDDTVITLADWYHLAAKVGQR
3*	FPLGADATLINGLGR
4*	TPGTTSADLAVIKVTQGK
5	YSFVLDASQPVDNYWIRANPPFGNVGFAGGINSAILR
6	SAGSSEYNYDNPVFR

[i] - * Contiguous peptides of the internal laccase fragment used in alignments

Table 3: Kinetics constants of *Pycnoporus sanguineus* CS 2 laccase

Substrate	Km (μM)	Vmax (μmol/min/ml)	kcat (s-1)	kcat /Km (s-1/M-1)
2,6-DMP	41	500	88	2.16 x 106
o-dianisidine	44	1111	197	4.49 x 106
ABTS	23	1250	221	9.38 x 106

Among these substrates this laccase showed preference for ABTS and this characteristic was consistent with most of fungal laccases [32, 41, 42]. Unexpectedly the substrate saturation graphics with SYR showed a sigmoidal-like behavior instead of the common hyperbolic one (not shown). This result could be explained considering a kinetic mechanism of positive cooperativity as that described for monomeric mnemonical enzymes [43], where a conformational change of interacting enzyme at the end of the first catalytic cycle, reacts more readily with a second substrate molecule than other free-enzyme. Other factor contributing to this result could be the presence of ethanol in routinely SYR assay affecting the substrate solubility and/or enzyme activity. Whether mechanistic on phenomenological, this observation must be taken into account in future works, considering the relevance of this substrate in laccase characterization.

Dye Decolorization

As revealed by the activity staining of native gels shown above, *P. sanguineus* CS 2 laccase was also able to promote the oxidative coupling between MBTH and DMAB. This reaction has been considered as indicative of the ability of some laccases to catalyze reactions requiring a higher redox potential, as in the case of the enzymatic decolorization of many synthetic dyes. The non-phenolic azo MR [44, 45] and diazo RB 5 dyes [46] have been used as models for studying the ability of laccases to degrade recalcitrant compounds (Figure 9). Although general consensus is that laccases require meditators for acting over these dyes, *P. sanguineus* CS 2 laccase showed the capability to decolorize directly both compounds, but with different ability. Decolorization of MR and RB 5 reached a level of 70 %, and 15% respectively, after 4 h at 25 °C.

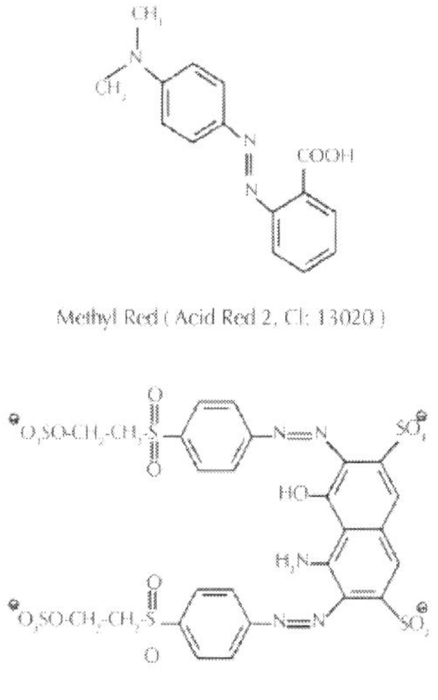

Methyl Red (Acid Red 2, Cl: 13020)

Reactive Black 5 (RB 5, Cl: 20505)

Figure 9: Chemical structure of the recalcitrant methyl red and reactive black 5 dyes used in this study.

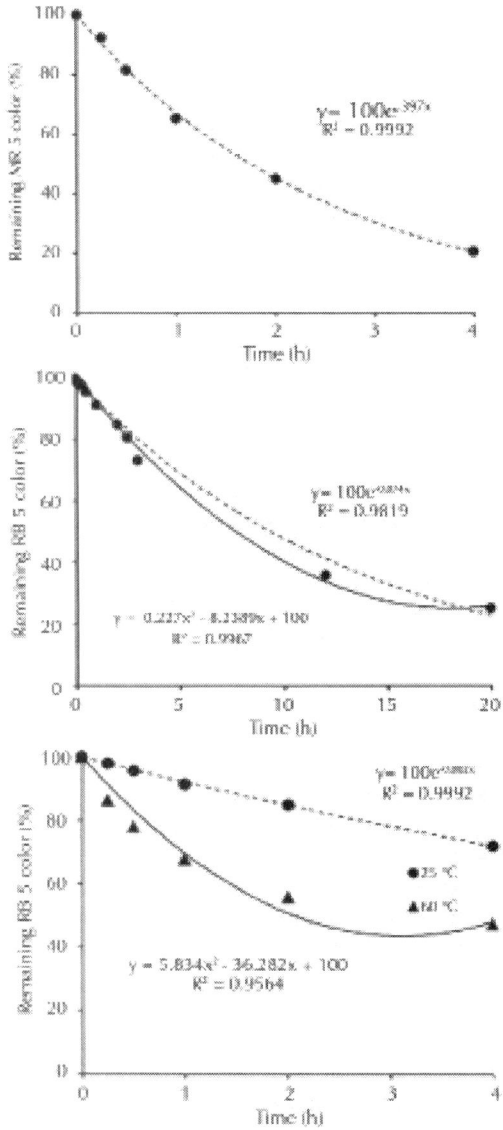

Figure 10: Decolorization of methyl red and reactive black 5 by *Pycnoporus sanguineus* CS2 laccase. Assays were performed by incubating 25 µM Methyl Red (upper panel) and 25 µM Reactive Black 5 (middle and lower panels) with laccase (5U/ml) in 0.2 mM citrate/phosphate buffer at pH 4.0 Aliquots were withdrawn from the assay mixture at the

indicated times and remaining color was determined as described in text. Lines showed the best data fittings corresponding to the exponential first-order (dashed) or polynomial second-order decay functions (continuous).

While *Ganoderma lucidum* [38], *Trametes trogii* [47] and *Lentinula edodes* [45] laccases were only able to decolorize RB 5 in the presence of mediators, a recent report state that three *Pycnoporus* laccases [13], were able to perform this decolorization in the absence of mediators, under similar conditions used in this work, with decolorization reaching from 29 to 45% after 52 h, at room temperature. The recalcitrance of RB 5 to laccase decolorization has been explained by its high redox potential or steric hindrances limiting accessibility of enzyme to –OH and –NH_2 groups in dye. As in this study native laccase attained around 70% decolorization after 20 h at room temperature, decolorization assays were performed at 60 °C taking advantage of its thermostability trying to overcome limiting factors. As expected, decolorization process was faster under the influence of temperature, reaching around 50% after 4 h, although it also seems to be limited faster (Figure 10). While MR decolorization fitted an exponential first order decay model, RB 5 decolorization changes rapidly from this behavior to fit a polynomial second order model. This effect could be related to several factors as an increased enzyme inactivation by endogenous generated reaction intermediates and/or dead-end transformation products. This relationship must be investigated in future work. Nonetheless these results illustrate the potential of the thermostable *Pycnoporus sanguineus* CS2 laccase for practical applications.

CONCLUSION AND FUTURE PROSPECTS

Its thermostability and ability for acting on high redox substrates and recalcitrant dyes, makes *Pycnoporus sanguineus* CS 2 laccase a good prospect for its application in industrial and environmental processes. This laccase could also be interesting as a model in studies associating structure-function of thermotolerant proteins from mesophilic microorganisms.

ACKNOWLEDGEMENTS

Authors thank the financial support provided by the Sistema de Fondos INNOVAPYME-CONACYT (Proyecto No. 139352). We also tank to Unidad de Proteómica, IBT-UNAM for assistance in peptide sequencing.

REFERENCES

1. Mayer AM, Staples RC. Laccase: New Functions for an Old Enzyme. Phytochemistry 2002;60 551-565.

2. Baldrian P. Fungal Laccases-Occurrence and Properties. FEMS Microbiological Reviews 2005;20 1-28.

3. Morozova OV, Shumakovich GP, Gorbacheva MA, Shleev SV, Yaropolov AI. "Blue" Laccases. Biochemistry (Moscow) 2007;72 1136–1412.

4. Thurston CF. The Structure of Fungal Laccases. Microbiology 1994 ;140 19-26.

5. Eggert C, Temp U, Eriksson K-EL. The Ligninolytic System of the White- rot Fungus Pycnoporus cinnabarinus: Purification and Characterization of the Laccase. Applied Environmental Microbiology 1996;62 1151–1158.

6. Leontievsky AA, Vares T, Lankinen P, Shergill JK, Pozdnyakona NN, Myasoedova NM. Blue and Yellow Laccases of Ligninolytic Fungi. FEMS Microbiological Letters 1997;156 9-14.

7. Palmieri G, Giardina, P, Bianco C, Sacloni A, Capasso A, Sannia G.A A Novel White Laccase from Pleurotus ostreatus. Journal of Biological Chemistry 1997;50 31301-31307.

8. Pointing SB. Feasibility of Bioremediation by White–rot Fungi. Applied Microbiology and Biotechnology 2001;57 20-33.

9. Reddy CA. The Potential of White-rot Fungi in the Treatment of Pollutants. Current Opinion in Biotechnology 1995;6 320-328.

10. Rodríguez-Couto S, Toca-Herrera JL. Laccases in the Textile Industry. Biotechnology and Molecular Biolology Reviews 2006a;1 115-120.

11. Rodríguez-Couto S, Toca-Herrera JL. Industrial and Biotechnological Applications of Laccases: A review. Biotechnology Advances 2006b;24 500-513.

12. Xu F, Berka RM, Wahleithner JA, Nelson BA, Shuster JR, Brown SH, Palmer AE, Solomon EI. Site-directed Mutations in Fungal Laccase: Effect on Redox Potential, Activity and pH Profile. Biochemistry Journal 1998;334(1) 63-70.

13. Uzan E, Nousiainen P, Balland V, Sipila J, Piumi F, Navarro D, Asther M, Record E, Lomascolo A. High Redox Potential Laccases from the Ligninolytic Fungi *Pycnoporus coccineus* and *Pycnoporus sanguineus* Suitable for White Biotechnology: from Gene Cloning to Enzyme Characterization and Applications. Journal of Applied Microbiology 2010;108 2199-2213.

14. Rodgers CJ, Blanford CF, Giddens SR, Skamnioti P, Armstrong FA Gurr SJ. Designer Laccases: A Vogue for High-potential Fungal Enzymes?. Trends in Biotechnology 2009;28(2) 63-72.

15. Jordaan J, Leukes WD. Isolation of a Thermostable Laccase with DMAB and MBTH Oxidative Coupling Activity from a Mesophilic White-rot Fungus. Enzyme and Microbial Technology 2003;33 212-219.

16. Dantán-González E, Vite-Vallejo O, Martínez-Anaya C, Méndez-Sánchez M, González MC, Palomares LA, Folch-Mallol J. Production of Two Novel Laccase Isoforms by a Thermotolerant Strain of *Pycnoporus sanguineus* Isolated from an Oil-polluted Tropical Habitat. International Microbiology 2008;11 163–169.

17. Pozdnyakova NN, Turkovskaya OV, Yudina EN Rodakiewicz-Nowak Y. Yellow Laccase from the Fungus *Pleurotus ostreatus* D1: Purification and Characterization. Applied Biochemistry and Microbiology 2006;42(1) 56-61.

18. Hernández-Luna CE, Gutiérrez-Soto G, Salcedo-Martínez SM. Screening for Decolorizing Basidiomycetes in Mexico. World Journal of Microbiology and Biotecnology 2008;24 465-473.

19. Ryvarden L. Genera of Polypores Nomenclature and Taxonomy. Synopsis fungorum Vol. 5. Fungiflora A/S Oslo; 1991.

20. Guzmán G. Identificación de los Hongos: Comestibles, Venenosos, Alucinantes y Destructores de la Madera: Editorial Limusa S.A. México; 1980.

21.   Abadulla E, Tzanov T, Costa S, Robra K, Gübitz G. Decolorization and Detoxification of Textile Dyes with a Laccase from *Trametes hirsuta*. Applied and Environmental Microbiology 2000;66 3357-3362.

22.   Tien M, Kirk TK. Lignin Peroxidase of *Phanerochaete chrysosporium*. Methods in Enzymology 1988;161 238-248.

23.   Kuan C, Johnson J, Tien M. Kinetic Analysis of Manganese Peroxidase. The Reaction with Manganese Complexes. Journal of Biological Chemistry 1993;268 20064–20070

24.   Pickard MA, Vandertol H, Roman R, Vazquez-Duhalt R. High Production of Ligninolytic Enzymes from White-rot Fungi in Cereal Bran Flakes Liquid Medium. Canadian Journal of Microbioliology 1999;45 627-631.

25.   Dhouib A, Hamza M, Zouari H, Mechichi T, Hmidi R, Labat M, Martinez MJ, Sayadi S. Screening for Ligninolytic Enzyme Production by Diverse Fungi from Tunisia. World Journal of Microbiology and Biotechnology 2005 ;21 1415-1423.

26.   Zouari-Mechichi H, Mechichi T, Dhouib A, Sayadi S, Martínez AT, Martínez MJ. Laccase Purification and Characterization from *Trametes trogii* isolated in Tunisia: Decolorization of Textile Dyes by the Purified Enzyme. Enzyme and Microbial Technology 2006;39 141-148.

27.   Laemmli U. Cleavage of Structural Proteins During the Assembly of the Head of the Bacteriophage T4. Nature 1970;227 680-685.

28.   Garfin DE. Methods in Enzymology. Academic Press Inc. 1990.

29.   Guzmán G. Los Hongos del Edén Quintana Roo. (*Introducción a la Micología Tropical de México*). INECOL y CONABIO, Xalapa ; 2003.

30.   Wang ZX, Caia YJ, Liao XR, Taoc GJ, Lia YY, Zhanga F, Zhang DB. Purification and Characterization of Two Thermostable Laccases with High Cold Adapted Characteristics from *Pycnoporus* sp. SYBC-L1. Process Biochemistry 2010;45 1720-1729.

31.   Litthauer D, Jansen van Vuuren M, van Tonder A, Wolfaardt FW. Purification and Kinetics of a Thermostable Laccase from *Pycnoporus sanguineus* (SCC 108). Enzyme Microbial Technology 2007;40 563–568.

32. Lu L, Zhao M, Zhang BB, Yu SY, Bian X-J, Wang W, Wang Y. Purification and Characterization of Laccase from *Pycnoporus sanguineus* and Decolorization of an Anthraquinone Dye by the Enzyme. Applied Microbiology and Biotechnology 2007;74 1232-1239.

33. Hernández-Fernaud JR, Marina A, González K, Vázquez J, Falcón MA. Production, Partial Characterization and Spectroscopic Study of the Extracellular Laccase Activity from *Fusarium proliferatum*. Applied Microbiology and Biotechnology 2006 ;70 212-221.

34. Madzak C, Mimmi MC, Caminade E, Brault A, Baumberger S, Briozzo P, Mougin C, Jolivalt C. Shifting the Optimal pH of Activity for a Laccase from the Fungus *Trametes versicolor* by Structure-based Mutagenesis. Protein Engineering 2006;334 63-70.

35. Hildén K, Hakala TK, Lundell T. Thermotolerant and Thermostable Laccases. Biotechnological Letters 2009;31 1117–1128.

36. Holm KA, Nielsen DM, Eriksen J. Automated Colorimetric Determination of Recombinant Fungal Laccase Activity in Fermentation Samples Using Syringaldazine as Chromogenic Substrate. Journal of Automated Chemistry 1998;20 199 –203.

37. Schneider K, Caspersen MB, Mondorf K, Halkier T, Skov LK, Ostergaard PR, Brown KM, Brown SH, Xu F. Characterization of a *Coprinus cinereus*Laccase. Enzyme and Microbial Technology 1999;25 502–508.

38. Murugesan K, Nam IH, Kim YM, Chang YS. Decolorization of Reactive Dyes by a Thermostable Laccase Produced by *Ganoderma lucidum* in Solid-state Culture. Enzyme and Microbial Technology 2007;40 1662-1672.

39. García TA, Santiago MF, Ulhoa CJ. Studies on the *Pycnoporus sanguineus* CCT-4518 Laccase Purified by Hydrophobic Interaction Chromatography. Applied Microbiology and Biotechnology 2007;75 311–318.

40. Antorini M, Herpoel-Gimbert I, Choinowski T, Sigoillot JC, Asther M, Winterhalter K, Piontek K. Purification, Crystallization and X-ray Diffraction Study of Fully Functional Laccases from Two Ligninolytic Fungi. Biochemistry and Biophysics Acta 2002;1594 (1) 109-114

41. Ko EM, Leem YE, Choi HT. Purification and Characterization of Laccase Isoenzymes from the White-rot Basidiomycete *Ganoderma lucidum*. Applied Microbiology and Biotechnology 2001;57 98-102.

42. Zang H, Zang Y, Huang F, Gao P, Chen J. Purification and Characterization of a Thermostable Laccase with Unique Oxidative Characteristics from *Tramentes hirsuta*. Biotechnology Letters 2009;31 837-843.

43. Ricard J, Noat G. Kinetic Co-operativity of Monomeric Mnemonical Enzymes 'The Significance of the Kinetic Hill-coefficient. European Journal of Biochemistry 1985;152 557-564.

44. Haibo Z, Yinglong Z, Feng H, Peiji G, Jiachuan C. Purification and Characterization of a Thermostable Laccase with Unique Oxidative Characteristics from *Trametes hirsuta*. Biotechnological Letters 2009;31 837-843.

45. Nagai M, Sato T, Watanabe H, Saito K, Kawata M, Enei H. Purification and Characterization of an Extracellular Laccase from the Edible Mushroom *Lentinula edodes*, and Decolorization of Chemically Different Dyes. Applied Microbiology and Biotechnology 2002;60 327-335.

46. Camarero S, Ibarra D, Martínez MJ, Martínez AT. Lignin-Derived compounds as Efficient Laccase Mediators for Decolorization of Different Types of Recalcitrant Dyes. Applied and Environmental Microbiology 2005;71(4) 1775-1784.

47. Zeng X, Cai Y, Liao X, Zeng X, Li W, Zhang D. Decolorization of Synthetic Dyes by Crude Laccase from a Newly Isolated *Trametes trogii* Strain Cultivated on Solid Agro-industrial Residue. Journal of Hazardous Materials 2011;189 517-525.

Structure and Photoluminescence of the Tio$_2$ Films Grown by Atomic Layer Deposition Using Tetrakis-dimethylamino Titanium and Ozone

Chunyan Jin[1], Ben Liu[1], Zhongxiang Lei[2], and Jiaming Sun[1]

[1]Key Laboratory of Weak Light Nonlinear Photonics, Ministry of Education, School of Physics, Nankai University, Weijin Road 94, Tianjin 300071, China

[2]Air Force Aviation University, Nanhu Road No. 2222, Changchun 130022, China

ABSTRACT

TiO$_2$ films were grown on silicon substrates by atomic layer deposition (ALD) using tetrakis-dimethylamino titanium and ozone. Amorphous

TiO$_2$ film was deposited at a low substrate temperature of 165°C, and anatase TiO$_2$ film was grown at 250°C. The amorphous TiO$_2$ film crystallizes to anatase TiO$_2$ phase with annealing temperature ranged from 300°C to 1,100°C in N$_2$atmosphere, while the anatase TiO$_2$ film transforms into rutile phase at a temperature of 1,000°C. Photoluminescence from anatase TiO$_2$ films contains a red band at 600 nm and a green band at around 515 nm. The red band exhibits a strong correlation with defects of the under-coordinated Ti^{3+} ions, and the green band shows a close relationship with the oxygen vacancies on (101) oriented anatase crystal surface. A blue shift of the photoluminescence spectra reveals that the defects of under-coordinated Ti^{3+} ions transform to surface oxygen vacancies in the anatase TiO$_2$film annealing at temperature from 800°C to 900°C in N$_2$ atmosphere.

PACS: 81.15.Gh, 72.20.Jv, 78.20.Ci.

BACKGROUND

TiO$_2$ has become a promising material in different applications for its large band gap [1], high refractive index [2],[3], high dielectric constant [4],[5], and highly active surface. In terms of photochemical properties, TiO$_2$ is used in decomposition of water into hydrogen and oxygen [6] and served as a photocatalyst in solar cells [7]. Degradation of organic molecules is another active research topic, such as purification of waste water [8], disinfection in public [9], self-cleaning coating [10], corrosion-protection [11], and actively suppressed impact on tumor cells of rats illuminated by near-UV [12]-[14]. In addition, TiO$_2$, as a semiconducting metal oxide, can be used as oxygen gas sensor to control the air/fuel mixture in car engines [15], [16]. The high dielectric constant broadens the applications of TiO$_2$ in electronics, such as capacitor and memory device. In our daily life, titanium dioxide pigment is almost used in every kind of paint because of its high refractive index. Moreover, pure TiO$_2$ is non-toxic and easy-dispersive, and it can be used in food additive [17], in cosmetic products, as well as in pharmaceuticals [18].

Among extensive applications using physical and surface chemical properties of TiO$_2$, the defects and the surface states of TiO$_2$, which depend strongly on material preparation technologies, play an important role in its electrical, chemical, as well as optical properties.

Therefore, selection of a well-controllable technology to engineer the defects in TiO$_2$ will be crucial for specific application. Almost all viable physical and chemical deposition technologies have been adopted to prepare TiO$_2$ thin films. Atomic layer deposition (ALD) has distinguished advantages over others for its precise thickness control, extremely conformal surface coating for nanostructures, large area uniformity, and low growth temperature [19]-[21]. Several precursors have been applied successfully for deposition of TiO$_2$ by ALD processes. The common precursor, TiCl$_4$, is a liquid with a moderate vapor pressure [22]-[27]. In the ALD process with H$_2$O/H$_2$O$_2$ as oxidant, the corrosive by-products of HCl and residual TiCl$_4$ are considered as a drawback. Same as Ti halide, TiI$_4$ can also be served as another precursor [28]-[30] with relative less corrosive, compared to TiCl$_4$. Recently, titanium alkoxides become promising precursors without corrosive halogen by-products, and research has been carried out on isopropoxide (Ti(O^{i}Pr)$_4$) and titanium ethoxide (Ti(OEt)$_4$). Although high purity thin films can be grown at 300°C, the decomposition of precursor leads to an undistinguished ALD temperature window. In addition, titanium isopropoxide can be adopted as precursor in theory, but significant decomposition occurs at lower deposition temperature than that of the titanium ethoxide. Since the bond energy of metal-halide is much stronger than that of the metal-nitrogen bond, metal amide compounds are expected to have much higher reactivity with H$_2$O, and therefore, tetrakis-(dimethylamino) titanium (TDMAT) and H$_2$O have been used for ALD processes[31],[32]. However, using H$_2$O as oxidant has two main disadvantages: the water vapor exposure on TiO$_2$ surface requires a very long purge time at the deposition temperature below 150°C [33], and the H$_2$O-based ALD process brings impurities, such as hydroxyl groups (−OH) in the films[34],[35]. The "dry" ALD process of TiO$_2$ films using TDMAT and ozone (O$_3$) may have more advantages, comparing to the TDMAT and H$_2$O process. Only a few reports have been published concerning the TDMAT and O$_3$ process [36],[37], and the study on controlling the transformation of structure and defects has not yet been done in ALD TiO$_2$ films. A comprehensive research on the thermal stability of the structure and defects in the ALD TiO$_2$ film is crucial for controlling its electric and optical properties for different applications.

In this study, TiO$_2$ films were deposited on silicon substrates by ALD technology using TDMAT and ozone process. The dependences of the growth rate, refractive index, and crystal structure and defects of the TiO_2 films on the growth temperatures are investigated in details by optical ellipsometry, X-ray diffraction (XRD), photoluminescence (PL), and X-ray photoelectron spectroscopy (XPS). Annealing processes were performed comparably on two as-grown TiO_2 films with amorphous and anatase phase structures, respectively. Thermal stability of the structure and defects in the as-grown TiO_2 films and those annealed at different temperatures were studied by PL spectroscopy in conjunction with XRD and XPS analysis. Amorphous, anatase, and rutile TiO_2 films were prepared at different ALD growth temperatures or by annealing at different temperatures. The PL spectra show a red band at 600 nm and a green band at around 515 nm from the defects in anatase TiO_2 films. It was shown that the red band has a strong correlation with the defects associated with under-coordinated Ti^{3+} ions and the green band is related to the oxygen vacancies on (101) surface of anatase TiO_2 films. The blue shift of the PL spectra indicates that the defects in anatase TiO_2 film undergo a transformation from under-coordinated Ti atoms to surface oxygen vacancies with increasing the annealing temperature from 800°C to 900°C in N_2 atmosphere.

METHODS

TiO_2 films were deposited on 4-in. (100) oriented n-type silicon wafers by a small chamber ALD system (Cambridge NanoTech Savannah 100, Cambridge NanoTech Inc., Cambridge, MA, USA) using TDMAT and O_3. The evaporation temperature of the TDMAT source was kept at 60°C, and the precursor delivery lines were heated at 150°C. O_3 was generated from high purity O_2 (99.999%) through an ozone generator with an O_2/O_3 flow of 500 sccm and O_3 concentration of 36 mg/L. High purity nitrogen gas (99.999%) was used as a carrying and purging gas with a flow rate of 20 sccm. Before the film deposition, the Si wafer was cleaned through the standard process of Radio Corporation of America (RCA), followed by a final cleaning in diluted HF solution. TiO_2 samples with 1,000 ALD cycles were deposited at different substrate temperatures varying from 75°C to 400°C. One TiO_2 deposition cycle consists of 0.5 s TDMAT pulse time, 5 s N_2 purge, 1.8 s O_2/O_3 pulse time, and 9 s N_2 purge, respectively. The thermal stability of the structures of ALD TiO_2 films was studied by annealing two as-grown ALD samples with different initial structures; one is an amorphous TiO$_2$ film grown at low substrate temperature of 165°C, and the other is an anatase TiO$_2$ film grown at 250°C. The annealing treatment was taken at different temperatures from 250°C to 1,150°C in a flowing N$_2$ atmosphere for 1 h.

The crystallinity of the TiO$_2$ films was characterized by XRD with Cu K$_\alpha$ radiation. The thickness and refractive index of the TiO$_2$ films on Si substrates were measured by an ellipsometer with a 632.8-nm He-Ne laser beam at an incident angle of 69.8°. The film growth per cycle was calculated by dividing the film thickness with the total number of ALD cycles. PL spectra of the TiO$_2$ films were measured at room temperature under the excitation of the 266-nm line of a pulsed diode pumped Q-switch solid state laser (CryLas DX-Q, CryLaS GmbH, Berlin, Germany). The PL signal was collected by a 1/2 meter monochromator and detected by a photomultiplier (model H7732-10, Hamamatsu Corporation, Shimokanzo, Iwata, Japan) connected to a computer-controlled Keithley 2010 multimeter (Keithley Instruments Inc., Cleveland, OH, USA). XPS measurement was performed in a Kratos Axis Ultra DLD spectrometer (Kratos Analytical Ltd, Britain). Monochromatized Al-Kα X-ray source (h_γ = 1,486.6 eV) was utilized to excite TiO$_2$ thin films. X-ray photoelectron spectra were measured from the surface of the TiO$_2$ samples annealed at 350°C, 600°C, 800°C, 850°C, and 1,000°C. For comparison, XPS was measured from the sample annealed at 1,000°C after removing 3 nm of the surface layer by Ar$^+$ ion sputtering. The Ar$^+$ ion sputtering was performed over an area of 2×2 mm^2, using an ion current of about 100 mA. The binding energy of each spectrum was calibrated by using the standard energy of carbon C1s peak at 284.6 eV.

RESULTS AND DISCUSSION

Figure 1 shows the dependences of the growth per cycle and the refractive index of the TiO$_2$ films on the growth temperature. Initially, the growth rate of the TiO$_2$ films decreases from 0.52 to 0.45 Å/cycle with increasing temperature from 75°C to 100°C, then a saturated growth window appears at the growth temperature from 100°C to 250°C, with a stable self-limiting growth rate of 0.46 Å/cycle. Further increasing the growth temperature above 300°C, the growth rate strongly increases.

The growth rate in Figure 1 is consistent with the results in ref. [37] in the same temperature range, which showed an ALD temperature window of 150°C to 225°C and a deposition rate of 0.44 ± 0.15 Å/cycle, respectively.

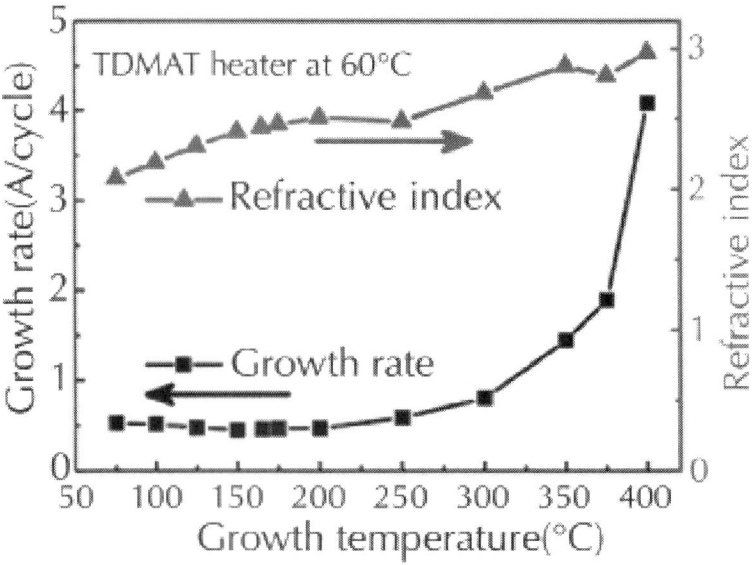

Figure 1: Dependences of the growth per cycle and refractive index on the growth temperature. The TiO$_2$ films were deposited on Si (100) substrates with 1,000 ALD cycles.

The dependence of the growth per cycle on the growth temperature in this O$_3$-based process is different from that of the TDMA and H$_2$O process reported in ref. [31],[32],[38]. In the H$_2$O-based process, where -OH groups are the reactive sites, the deposition rate decreases with increasing the growth temperature from 80°C to 350°C. The growth rate of TiO$_2$ at growth temperature below 150°C could be strongly influenced by the purging time of H$_2$O vapor. The reported results on the growth rate per cycle are controversial using TDMAT and H$_2$O process, depending on the ALD systems used by different groups. Lim and Kim observed a narrow ALD window between 120°C and 150°C in the TDMAT and H$_2$O process [31], whereas other reports showed a decrease of the growth rate with increasing temperature from 150°C up to 350°C, without distinguished saturated growth temperature window [32],[39]. This is probably due to the insufficient evacuation of the residual H$_2$O vapor in the growth chamber [40]. The decrease of the growth rate in the H$_2$O-based process with increasing the growth temperature was probably caused by the strong thermal desorption of the intermediate product mediated by -OH group adsorption on the surface, as proposed in ref. [32]. The possible chemical reaction of the surface species with -OH groups is:

$$TiO_2\text{-}O\text{-}Ti\,[N(CH_3)_2]_3{}^* + TiO_2\text{-}OH^* \rightarrow TiO_2\text{-}O^*\text{-}TiO_2 + HOTi[N(CH_3)_2]_3\uparrow.$$

On the contrary, in the O$_3$-based process, the desorption of intermediate products is suppressed without the surface adsorption of the -OH groups from H$_2$O vapor; therefore, a wider ALD saturated growth temperature window from 100°C to 250°C was observed, and the growth rate shows a strong increase of from 0.58 to 4.08 Å/cycle with increasing the growth temperature from 250°C to 400°C. The strong increase of the growth rate is due to the chemical vapor deposition (CVD) process which is related to the strong thermal decomposition of the TDMAT precursor [41],[42] at temperature above 250°C.

Figure 2 shows the XRD patterns of the TiO$_2$ films deposited at different growth temperatures from 175°C to 400°C. Initially, the films deposited at temperatures below 175°C are amorphous. With increasing growth temperature from 200°C to 250°C, the films show anatase crystal phase, with the (101) and (200) peaks in the diffraction patterns. The intensity of the anatase (101) peak reaches a maximum at the growth temperature of 250°C and then decreases dramatically to 300°C, with an emergence of a weak (110) peak from rutile TiO$_2$. At growth temperature above 250°C, the growth mode of the films changes to fast CVD mode, the fast deposition rate causes a strong degradation of the crystallinity of the TiO$_2$ film, as shown by the decrease of the diffraction peaks in the XRD patterns at 300°C to 400°C. Despite of this, very weak (101) peak from anatase TiO$_2$ and (110) peak from rutile TiO$_2$ are observed in the XRD patterns, indicating the formation of a small among of rutile TiO$_2$ in the films. As it was reported that rutile TiO$_2$ is the stablest and densest structure of TiO$_2$ with a mass density of 4.25 g/cm^3, while the anatase TiO$_2$ is a metastable and less dense structure, with a smaller density of 3.894 g/cm^3[43]. The increasing tendency of the refractive index from 2.07 to 2.97 in Figure 1, which can be interpreted by the structure change in the films with increasing the growth temperature, is probably due to the film densification with the change from amorphous to anatase as well as the formation of rutile phase[44].

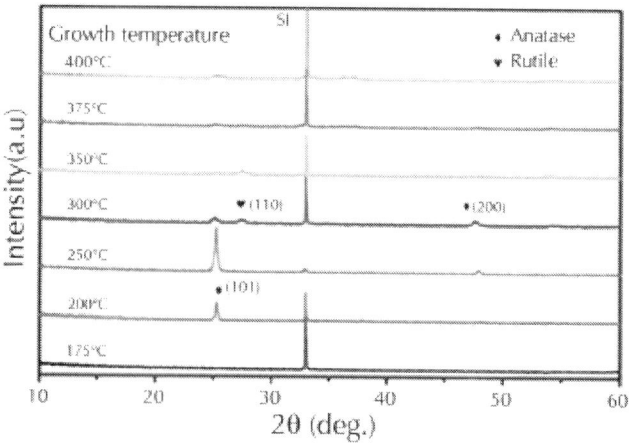

Figure 2: XRD patterns of the TiO$_2$films deposited at different temperatures.

The change of defects in the TiO$_2$ films was characterized by measuring the room-temperature PL spectra at different growth temperatures. As it was shown in Figure 3, no PL emission was detected from the amorphous TiO$_2$ films grown below 175°C. A green PL band at around 500 nm was observed from the TiO$_2$ films with anatase phase grown at temperatures from 200°C to 300°C. In order to study the correlation between the PL and the structure change in the films, the dependences of PL peak intensity and the intensity of (101) anatase peak from the XRD patterns on the growth temperature are plotted together in Figure 4. The PL intensity increases with increasing substrate temperature, reaches a maximum at 250°C, and then decreases strongly at higher growth temperature, which is similar to the growth temperature dependence of the (101) anatase peak intensity in the XRD patterns. This similarity indicates that the defects related to the green PL band are probably located on the (101) oriented surface of the anatase TiO$_2$ crystals. Finally, the strong quenching of the PL intensity at growth temperature over 250°C is probably due to the degradation of the anatase crystallinity by CVD, which causes an increase of the non-radiative recombination centers in the films.

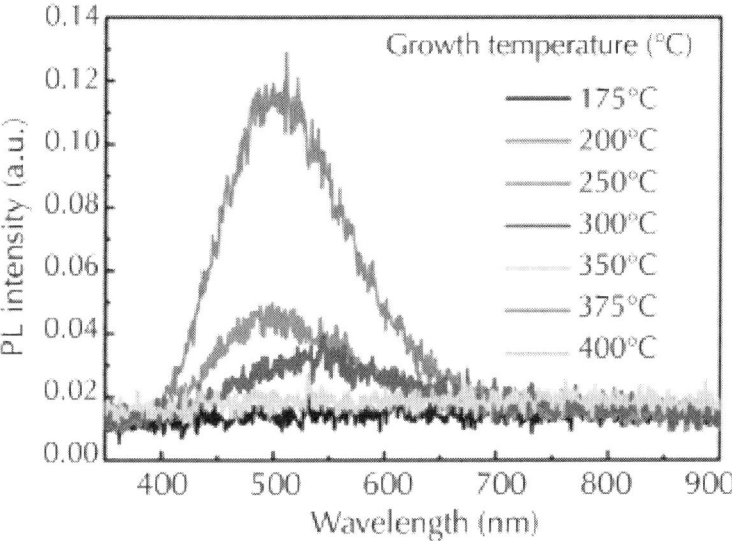

Figure 3: Room-temperature PL spectra from TiO$_2$ films grown at different temperatures. The spectra were taken under excitation of a 266 nm laser.

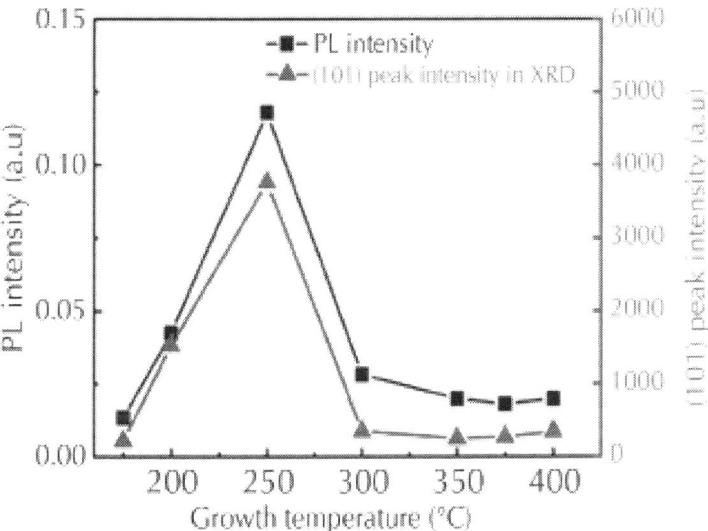

Figure 4: Dependences of the PL intensity and the anatase (101) peak intensity in XRD patterns on the growth temperature.

The thermal stability of the structures in ALD TiO$_2$ films was studied by annealing two as-grown samples with different initial structures in N$_2$ atmosphere; one is an amorphous TiO$_2$ film grown at low substrate temperature of 165°C, and the other is an anatase TiO$_2$ film grown at 250°C. Figure 5 shows the XRD patterns from the TiO$_2$ films after annealing the as-grown amorphous sample at different temperatures in N$_2$ atmosphere for 1 h. Initially, the films are still amorphous at lower annealing temperature below 250°C, then they crystallize to anatase TiO$_2$ in a very wide annealing temperature range from 300°C to 1,100°C, as shown in the emergence of the (101), (004), and (200) peaks of anatase TiO$_2$ in the XRD patterns in Figure 5.

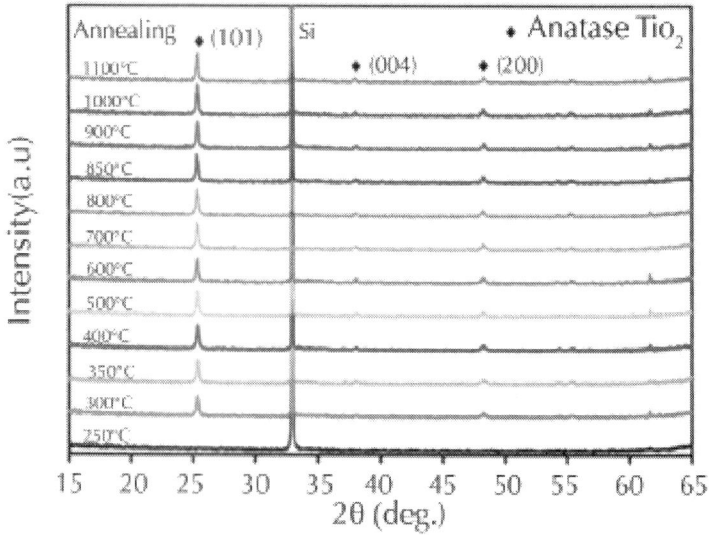

Figure 5: XRD patterns of the TiO2 films annealed at different temperatures in N$_2$ atmosphere. The as-grown TiO$_2$ film is amorphous deposited at a substrate temperature of 165°C.

Thermal stability of the defects in the annealed TiO$_2$ films in Figure 5 was characterized by the PL spectra in Figure 6. The PL spectra can be decomposed into a red band at 600 nm and a green band at 515 nm. The insert shows the areal percentages of the red emission and the green band, which are derived from multiple-peak Gaussian fitting of the PL spectra, respectively. In conjunction with the XRD patterns in Figure 5, no PL emission is observed from the amorphous TiO_2 films at annealing temperature lower than 250°C. The red PL band was observed from the defects in the anatase TiO_2 films, which were obtained by annealing the as-grown amorphous films at annealing temperatures from 300°C to 800°C. The intensity of the red band increases to a maximum for annealing temperature from 300 to 800°C, and then, it decreases strongly with increasing temperature. For annealing temperature from 800°C to 900°C, the green band appears with the decrease of the red band and the PL spectra undergo a crossover from the red band-dominated emission to the green band-dominated emission as shown in the insert. At even higher annealing temperatures from 900°C to 1,100°C, the PL spectra are dominated by the green band with a saturated intensity.

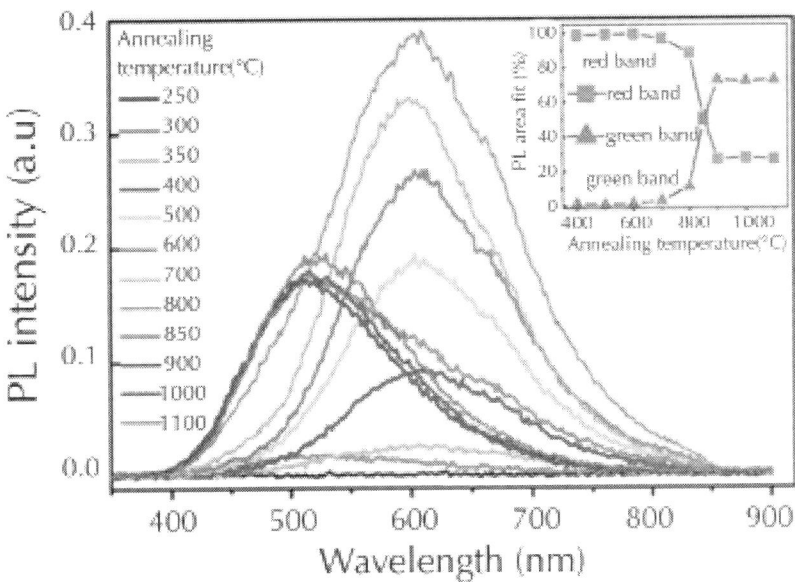

Figure 6: Room-temperature PL spectra of the TiO_2 films annealed at different temperatures in N2atmosphere. The as-grown TiO_2 film is amorphous deposited at a substrate temperature of 165°C. The insert is the areal percentages of the red emission band at 600 nm and the green emission band at 515 nm, which are derived from multiple-peak Gaussian fitting of the PL spectra, respectively.

Figure 7 is the XRD patterns from the TiO_2 films after annealing at different temperatures in N_2atmosphere for 1 h, in which the as-grown TiO$_2$ film was initially in anatase phase deposited at a substrate temperature of 250°C. Initially, the annealed TiO$_2$ films still keep anatase phase in a wide annealing temperature range from 400°C to 900°C, and then, a clear transition from anatase to rutile phase was observed in the annealing temperature range from 950°C to 1,000°C. Finally, the anatase films change to rutile TiO$_2$ phase at elevated annealing temperatures above 1,000°C.

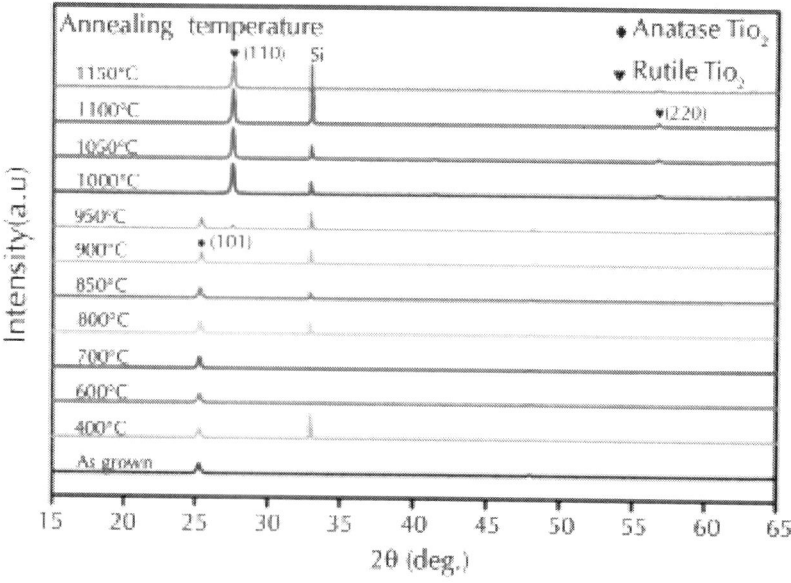

Figure 7: XRD patterns of TiO$_2$ films after annealing at different temperatures in N2 atmosphere. The as-grown TiO$_2$ film is anatase deposited at a substrate temperature of 250°C.

Figure 8 shows the room-temperature PL spectra for the TiO$_2$ samples in Figure 7 annealed at different temperatures, in which the as-grown film is in anatase TiO$_2$ phase. The broadband PL emission from the anatase TiO$_2$ films can be divided into two components with the red peak centered at 600 nm and the green peak centered at 515 nm. Gaussian fitting of the PL spectra was performed using the parameters of the red and green peaks derived from Figure 6. The insert shows the areal percentage of the red and the green components, respectively. The PL spectrum from the as-grown anatase TiO$_2$ film contains 24% of the red component and 76% of the green component. With increasing annealing temperature from 400°C to 700°C, the PL intensity of the sample increases slightly, the red component of the spectra increases from 24% to 85%, while the green component reduces from 76% to 15%; therefore, the PL spectra exhibit a red shift. For increasing the annealing temperature from 700°C to 850°C, a strong increase of the PL intensity was observed, the red component decreases from 85% to 47%, while the green component rises from 15% to 53% in the spectra, the relative increase of the green component causes a blue shift of the PL peak. Further increasing the annealing temperature from 850°C to 1,000°C, the PL peaks in the visible spectral range decrease dramatically due to the transition from the anatase to rutile TiO_2 phase. Finally, the PL spectra show a near-infrared peak at 820 nm from the defects in rutile TiO_2 formed at elevated annealing temperatures above 1,000°C.

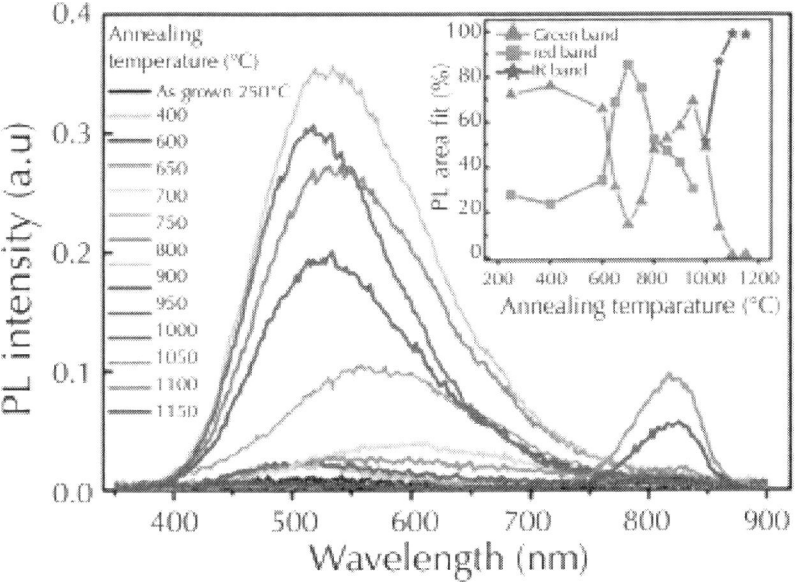

Figure 8: Room-temperature PL spectra of the TiO_2 films annealed at different temperatures in N2atmosphere. The as-grown TiO_2 film is anatase deposited at a substrate temperature of 250°C. The insert is the areal percentages of the red emission band at 615 nm, the green emission band at 510 nm, and the near-infrared peak at 820 nm, which are derived from multiple-peak Gaussian fitting of the PL spectra, respectively.

XPS spectroscopy was studied after annealing the as-grown amorphous TiO_2 films at different temperatures. Figure 9 shows the XPS peaks of Ti $2p_{3/2}$ and Ti $2p_{1/2}$ from the TiO_2 film annealed at 1,000°C after removing 3-nm surface layer by Ar^+ ion sputtering (a), the film annealed at 1,000°C without Ar^+ ion sputtering (b), and the film annealed at 800°C without Ar^+ ion sputtering (c). The binding energies of Ti $2p_{3/2}$ and Ti $2p_{1/2}$ peaks of Ti^{4+} ions in the sample annealed at 1,000°C are located at about 458.75 and 464.48 eV, respectively. After the removal of 3-nm surface layer by Ar^+ ion sputtering, the Ti $2p_{3/2}$ peak shifts to lower energy at 458.54 eV and a shoulder peak at a lower energy of 457.0 eV appears. Multiple-peak Gaussian fitting of the spectrum indicates that the peak at lower energy belongs to the valence state of Ti^{3+} ions in the TiO_2 film, which are formed by Ar^+ ion sputtering, as reported in ref. [32],[45],[46]. The presence of the Ti^{3+} states in the films causes a small shift of the $2p_{3/2}$ peak of Ti^{4+} ions to lower energy compared to the un-sputtered one in Figure 9b. As a consequence, comparing the $2p_{3/2}$ peak of Ti^{4+} in the sample annealed at 800°C (c) with the one annealed at 1,000°C (b), a slight shift of Ti $2p_{3/2}$ peak from 458.75 to lower energy of 458.46 eV was also observed for the sample annealed at 800°C (c). This suggests that a small amount of trivalent Ti^{3+} ions exist in the sample annealed at 800°C. The relative concentration of the Ti^{3+} ions with respect to the total Ti atoms in the annealed anatase TiO_2 films can be calculated from the integrated intensity of the $2p_{3/2}$ peak of Ti^{3+} ions (red dashed peak), which was derived by multiple-peak Gaussian fitting of the Ti $2p_{3/2}$ peak of the XPS, as shown by the dashed curves in Figure 9.

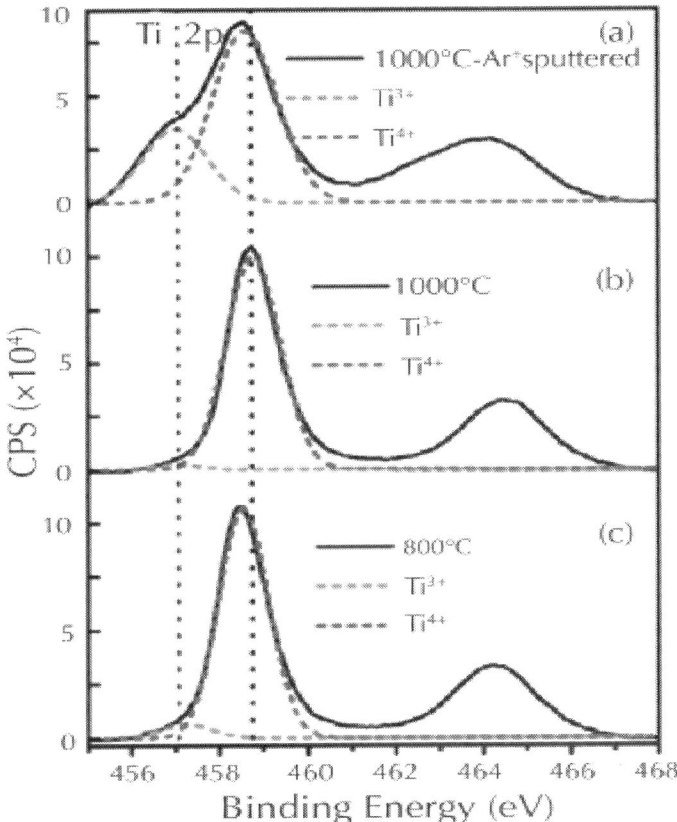

Figure 9: XPS spectra of Ti 2p states from the TiO₂films. Samples were annealed at 1,000°C after removing 3-nm surface layer by Ar⁺ ion sputtering (a); annealed at 1,000°C without Ar⁺ ion sputtering (b); and annealed at 800°C without Ar⁺ ion sputtering (c). The dashed curves are multiple-peak Gaussian fitting of the Ti $2p_{3/2}$ peak with two components from the valence states of Ti⁴⁺ (blue) and Ti³⁺ (red).

In order to clarify the correlation between the photoluminescence and the defects related to the under-coordinated Ti³⁺ ions in the annealed TiO₂ films, the integrated PL intensity of the red and green peaks as well as the percentage of Ti³⁺ in the films are plotted together as functions of the annealing temperature in Figure 10, in which the integrated PL intensity of the red and green peaks is derived by multiple-peak Gaussian fitting of the PL spectra in Figure 6. The integrated intensity of the green band was low at annealing temperature below 700°C, it increases from 700°C to 900°C, and then saturated at annealing temperature above 900°C. No obvious correlation was observed between the PL intensity of the green band and the Ti^{3+} ion concentration. The dependence of the integrated PL intensity of the red band on the annealing temperature shows a thermal behavior quite similar to the change of the Ti^{3+} ion concentration. Both of them increased with increasing annealing temperature from 300°C to 800°C, after reaches a maximum at 800°C and then decreases dramatically at annealing temperature varied from 800°C to 900°C. This similarity suggests that the red band may have a strong correlation with the defects associated with the under-coordinated Ti^{3+} ions in anatase TiO_2.

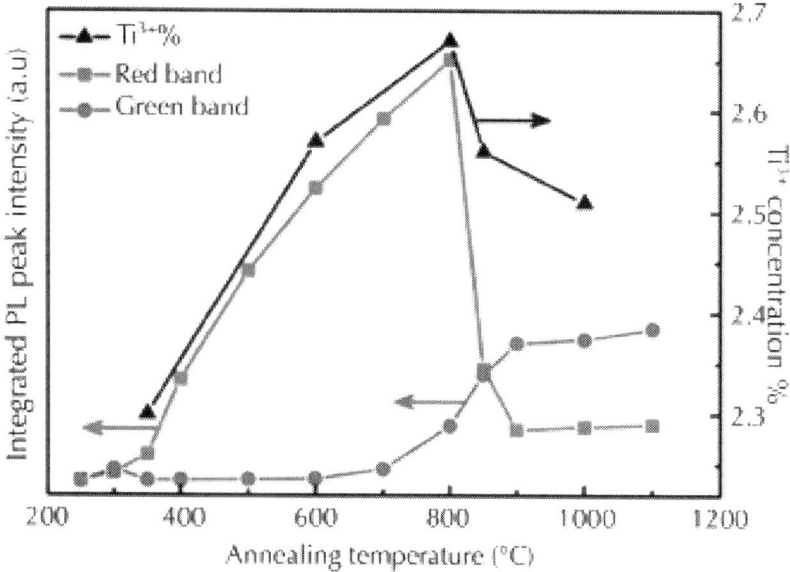

Figure 10: Calculated concentration of Ti^{3+} and the integrated PL intensity. The calculated concentration of Ti^{3+} from XPS analysis (black triangles) and the integrated PL intensity from the red (red squares) and green (green dots) bands is derived from the Gaussian fitting of the XPS and PL spectra of the TiO_2 films after annealing at different temperatures. The solid lines are for a guide of eyes.

Olson et al. [47] calculated the energies for creation of various defects, such as oxygen vacancy and Ti^{4+} and Ti^{3+} interstitials in TiO_2. The formation energy is 24.10 eV (E_1) for oxygen vacancy, which is higher than that for Ti^{3+} interstitial ($E_2 = -40.5$ eV) and Ti^{4+} interstitial ($E_3 = -77.23$ eV). The negative values of E_2 and E_3 indicate that the formation of the defects of under-coordinated Ti^{3+} is energetically favorable. Therefore, defects of Ti^{4+} interstitials and under-coordinated Ti^{3+}may form in higher priority at low annealing temperature during the crystallization of anatase crystal TiO$_2$ from the as-grown amorphous film. Thus, the red band associated with electron traps of under-coordinated Ti^{3+} dominates the PL spectra at low annealing temperature in Figure 6. The defects of under-coordinated Ti^{3+} can be created with the removal of oxygen atoms by annealing in an inert N$_2$ atmosphere or by Ar$^+$ ion sputtering. The removal of oxygen atom can create lone pair electrons to two neighboring Ti^{4+}, and then, electrons will reduce Ti^{4+} to Ti^{3+}. This is confirmed by our XPS study in Figure 9.

Concerning the green PL band from the anatase TiO$_2$ films, it dominates the PL spectra from the anatase TiO$_2$ films grown at 250°C or the anatase TiO$_2$ films which have undergone an annealing process at a high temperature above 850°C. This suggests that origin of the green band is probably from the relative stable surface oxygen vacancies on anatase TiO$_2$ films. The strong correlation of the green band PL intensity with the intensity of (101) peak in XRD patterns of ALD grown TiO$_2$ films in Figure 4 reveals that the green PL peak is related to the defects located on (101) surface in the anatase phase. Shi et al. and Mercado et al. [48],[49] studied the PL emission from TiO$_2$ nanocrystals, and they also draw the same conclusion that the green band emission is related to oxygen vacancies on exposed (101) surfaces of anatase TiO$_2$ nanocrystals. Since the Ti^{3+} ions are unstable, as shown by the dependence of the Ti^{3+} ion density on the annealing temperature in Figure 10, the defects of under-coordinated Ti^{3+} ions can be annealed out, immigrate, and transform into stable surface oxygen vacancies on the anatase TiO$_2$ films at high annealing temperature from 800°C to 900°C. Figure 6 shows a transition from the red band-dominated PL to the green band-dominated emission with increasing the annealing temperature from 800°C to 900°C. This reveals that some of the under-coordinated Ti^{3+} ions can transform into stable surface oxygen vacancies at high annealing temperature. This causes an increase of the intensity of the green PL band at annealing temperature from 800°C to 900°C. Since only the stable surface oxygen vacancies are preserved at elevated temperature [50]. Finally, the PL spectra are dominated by the green band with a saturation intensity at annealing temperatures above

900°C, as it is shown in Figures 6 and 10. This is also in accordance with the conclusion in ref. [51] that the green-emitting defects are oxygen vacancies located on the surface of anatase TiO$_2$ films.

From the results of this study and the comprehensive study of the luminescent defects in TiO$_2$ nanocrystals in ref. [48]-[51], the proposed model for PL in the ALD TiO$_2$ films is illustrated in Figure 11. After electrons are excited from the valence band to the conduction band of TiO$_2$, some electrons are captured by the electron traps associated with under-coordinated Ti atoms, which located at 0.7 to 1.6 eV below the conduction band edge. Radiative recombination of the electrons trapped around under-coordinated Ti atoms with the holes in the valence band contributes to the red band at around 600 to 620 nm. In addition, the green band at around 500 to 520 nm may be from the radiative recombination of free electrons with holes trapped around surface oxygen vacancies, which were located at 0.7 to 1.4 eV above the valence band edge. In addition, the near-infrared emission band at around 820 nm is from the defects in rutile TiO$_2$, which are related to the radiative recombination of electrons in conduction band with hole traps on the (110) and (1$\bar{1}$0) facets of oxygen vacancies [49].

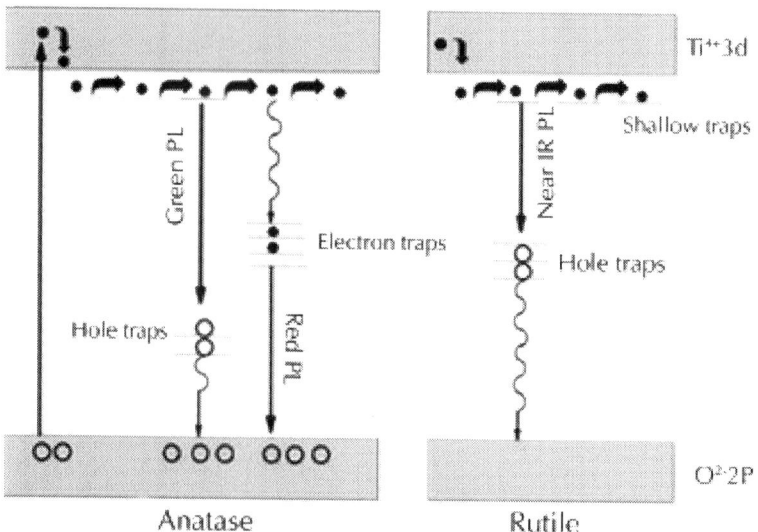

Figure 11: Models for photoluminescence from the electronic transitions of trap states in anatase and rutile TiO$_2$.

CONCLUSIONS

TiO$_2$ films were grown on silicon substrates by ALD using TDMAT precursor and ozone. A wide ALD growth window was observed between 100°C and 250°C with a self-limiting saturated growth rate of 0.46 Å/cycle. The film is amorphous at the growth temperatures of 165°C and then exhibits anatase crystal phase at the growth temperatures of 250°C. The initial amorphous TiO$_2$ sample crystallizes to anatase phase with annealing temperature from 300°C up to 1,100°C, while the initial anatase TiO$_2$ film transfers to rutile phase at elevated annealing temperature above 950°C. Photoluminescence spectra from the defects in the anatase TiO$_2$ films contain a red band at 600 nm and a green band at 515 nm. XPS and XRD studies indicate that the red band has a strong correlation with the defects of under-coordinated Ti^{3+} ions and the green band is related to the oxygen vacancies located on the (101) surface of the anatase TiO$_2$ films. The blue shift of the photoluminescence reveals that the defects in anatase TiO$_2$ film undergo a transition from under-coordinated Ti atoms to surface oxygen vacancies with increasing annealing temperature from 800°C to 900°C in N$_2$ atmosphere.

AUTHORS' CONTRIBUTIONS

CYJ performed the data analysis and drafted the manuscript. BL performed the growth of the samples, taking the analysis of XRD patterns and PL spectra. ZXL perform the technical support of the PL and XPS analysis. JMS carried out the design and the preparation of the study, supervised the work, and critically read the manuscript. All authors read and approved the final manuscript.

ACKNOWLEDGEMENTS

One of the authors would like to acknowledge Mrs. C.M. Shi for assisting in X-ray diffraction analysis. This work was supported by the Chinese "973" project (no. 2013CB632102) and National Natural Science Foundation of China NSFC (nos. 61275056 and 60977036).

REFERENCES

1.   Tang H, Prasad K, Sanjinès R, Schmid PE, Lévy F: Electrical and optical properties of TiO$_2$ anatase thin films. *J Appl Phys*. 1994, 75:2042.

2.   Chao S, Wang WH, Lee CC: Low-loss dielectric mirror with ion-beam-sputtered TiO$_2$-SiO$_2$ mixed films. *Appl Opt*. 2001, 40:2117-82.

3.   Yokogawa T, Yoshii S, Tsujimura A, Sasai Y, Merz J: Electrically pumped CdZnSe/ZnSe blue-green vertical-cavity surface-emitting lasers. *J J Appl Phys, Part 2: Letters* 1995, 34:L751-3.

4.   Fukuda H, Namioka S, Miura M, Ishikawa Y, Yoshino M, Nomura S: Structural and electrical properties of crystalline TiO$_2$ thin films formed by metalorganic decomposition. *J J Appl Phy*. 1999, 38:6034.

5.   Stephen AC, Wang XC, Hsieh MT, Kim HS, Gladfelter WL, Yan JH: MOSFET transistors fabricated with high permittivity TiO$_2$ dielectrics. *IEEE Trans Electron Devices*. 1997, 44:104-9.

6.   Fujishima A, Honda K: Photolysis-decomposition of water at the surface of an irradiated semiconductor. *Nature*. 1972, 238:37-8.

7.   O'Regan B, Gratzel M: A low-cost, high-efficiency solar cell based on dye-sensitized colloidal TiO$_2$ films. *Nature*. 1991, 353:737-40.

8.   Mills A, Davies RH, Worsley D: Water purification by semiconductor photocatalysis. *Chem Soc Rev* 1993, 22:417-25.

9.   Maness PC, Smolinski S, Blake DM, Huang Z, Wolfrum EJ, Jacoby WA: Bactericidal activity of photocatalytic TiO$_2$ reaction: toward an understanding of its killing mechanism. *Appl Environ Microbiol*. 1999, 65:4094-8.

10.   Paz Y, Luo Z, Rabenberg L, Heller A: Photooxidative self-cleaning transparent titanium dioxide films on glass. *Mater Res Soc*. 1995, 10:11.

11.   Poulios I, Spathis P, Grigoriadou A, Delidou K, Tsoumparis P: Protection of marbles against corrosion and microbial corrosion with TiO$_2$ coatings. *J Environ Sci Health Part A*. 1999, 34:1455-71.

12.   Cai R, Hashimoto K, Itoh K, Kubota Y, Fujishima A: Photokilling

of malignant cells with ultrafine TiO$_2$ powder. *Chem Soc Japan.* 1991, 64:4.

13. Lazar MA, Varghese S, Nair SS: Photocatalytic water treatment by titanium dioxide: recent updates. *Catalysts.* 2012, 2:572-601.

14. Sakai H, Baba R, Hashimoto K, Kubota Y, Fujishima A: Selective killing of a single cancerous T24 cell with TiO$_2$ semiconducting microelectrode under irradiation. *Chem Lett.* 1995, 24:185-6.

15. Dutta PK, Ginwalla A, Hogg B, Patton BR, Chwieroth B, Liang Z, *et al.*: Interaction of carbon monoxide with anatase surfaces at high temperatures: optimization of a carbon monoxide sensor. *J Phys Chem B.* 1999, 103:4412-22.

16. Xu Y, Yao K, Zhou X, Cao Q: Platinum-titania oxygen sensors and their sensing mechanisms. *Sens Actuators B.* 1993, 14:492-4.

17. Phillips LG, Barbano DM: The influence of fat substitutes based on protein and titanium dioxide on the sensory properties of lowfat milks. *J Dairy Sci.* 1997, 80:2726-31.

18. Tryk DA, Fujishima A, Honda K: Recent topics in photoelectrochemistry: achievements and future prospects. *Electrochim Acta.* 2000, 45:2363-76.

19. Montereali RM: Point defects in thin insulating films of lithium fluoride for optical microsystems. In *Handbook of thin film materials volume 3 ferroelectric and dielectric thin films.* Edited by Nalwa HS. Academic Press, San Diego San Francisco New York Boston London Sydney Tokyo; 2002:399-431.

20. Leskela M, Ritala M: Atomic layer deposition (ALD): from precursors to thin film structures. *Thin Solid Films.* 2002, 409:138-46.

21. Sneh O, Clark-Phelps RB, Londergan AR, Winkler J, Seidel TE: Thin film atomic layer deposition equipment for semiconductor processing. *Thin Solid Films.* 2002, 402:248-61.

22. Ritala M, Leskelä M, Nykänen E, Soininen P, Niinistö L: Growth of titanium dioxide thin films by atomic layer epitaxy. *Thin Solid Films.* 1993, 225:288-95.

23. Ritala M, Leskelä M, Johansson L-S, Niinistö L: Atomic force microscopy study of titanium dioxide thin films grown by atomic layer epitaxy. *Thin Solid Films.* 1993, 228:32-5.

24. Aarika J, Aidla A, Sammelselgb V, Siimon H, Uustare T: Control of thin film structure by reactant pressure in atomic layer deposition of TiO_2.

25. *J Cryst Growth.* 1996, 169:496-502.

26. Aarik J, Aidla A, Uustare T, Sammelselg V: Morphology and structure of TiO_2 thin films grown by atomic layer deposition.

27. *J Cryst Growth.* 1995, 148:268-75.

28. Aarik J, Aidla A, Kiisler AA, Uustare T, Sammelselg V: Effect of crystal structure on optical properties of TiO_2 films grown by atomic layer deposition. *Thin Solid Films.* 1997, 305:270-3.

29. Kumagai H, Matsumoto M, Toyoda K, Obara M, Suzuki M: Fabrication of titanium oxide thin films by controlled growth with sequential surface chemical reactions. *Thin Solid Films.* 1995, 263:47-53.

30. Kukli K, Ritala M, Schuisky M, Leskelä M, Sajavaara T, Keinonen J, et al.: Atomic layer deposition of titanium oxide from TiI_4 and H_2O_2. *Chem Vap Deposition.* 2000, 6:303-10.

31. Schuisky M, Aarik J, Kukli K, Aidla A, Hårsta A: Atomic layer deposition of thin films using O_2 as oxygen source. *Langmuir.* 2001, 17:5508-12.

32. Aarik J, Aidla A, Uustare T, Kukli K, Sammelselg V, Ritala M, et al.: Atomic layer deposition of TiO_2 thin films from TiI_4 and H_2O. *Appl Surf Sci.* 2002, 193:277-86.

33. Lim GT, Kim D-H: Characteristics of TiOx films prepared by chemical vapor deposition using tetrakis-dimethyl-amido-titanium and water. *Thin Solid Films.* 2006, 498:254-8.

34. Xie Q, Jiang YL, Detavernier C, Deduytsche D, Van Meirhaeghe RL, Ru GP, et al.: Atomic layer deposition of TiO_2 from tetrakis-dimethyl-amido titanium or Ti isopropoxide precursors and H_2O. *J Appl Phys.* 2007, 102:083521.

35. Dennis M, Hausmann EK, Becker J, Gordon RG: Atomic layer deposition of hafnium and zirconium oxides using metal amide precursors. *Chem Mater.* 2002, 14:4350-8.

36. Cleveland ER, Henn-Lecordier L, Rubloff GW: Role of surface intermediates in enhanced, uniform growth rates of TiO_2 atomic layer deposition thin films using titanium tetraisopropoxide and ozone. *J Vac Sci Technol A* 2012, 30:01A150.

37. Kurtz RL, Stock-Bauer R, Msdey TE, Román E, De Segovia J: Synchrotron radiation studies of H$_2$O adsorption on TiO$_2$(110). *Surf Sci.* 1989, 218:178-200.

38. Katamreddy R, Omarjee V, Feist B, Dussarrat C: Ti source precursors for atomic layer deposition of TiO$_2$, STO and BST. *ECS Trans.* 2008, 16:113-22.

39. Kim YW, Kim DH: Atomic layer deposition of TiO$_2$ from tetrakis-dimethylamido-titanium and ozone. *Korean J Chem Eng.* 2012, 29:969-73.

40. Rai VR, Agarwal S: Surface reaction mechanisms during plasma-assisted atomic layer deposition of titanium dioxide. *Phys Chem C.* 2009, 113:12962-5.

41. Nam T, Kim JM, Kim MK, Kim H, Kim WH: Low-temperature atomic layer deposition of TiO$_2$, Al$_2$O$_3$, and ZnO thin films. *J Korean Phy Soc.* 2011, 59:452-7.

42. Xie Q, Musschoot J, Deduytsche D, Van Meirhaeghe RL, Detavernier C, Van den Berghe S, *et al.*: Growth kinetics and crystallization behavior of TiO$_2$ films prepared by plasma enhanced atomic layer deposition. *J Electrochem Soc.* 2008, 155:H688.

43. Elam JW, Schuisky M, Ferguson JD, George SM: Surface chemistry and film growth during TiN atomic layer deposition using TDMAT and NH3. *Thin Solid Films.* 2003, 436:145-56.

44. Norton ET, Amato-Wierda C Jr: Kinetic and mechanistic studies of the thermal decomposition of Ti(N(CH$_3$)$_2$)$_4$ during chemical vapor deposition by in situ molecular beam mass spectrometry. *Chem Mater.* 2001, 13:4655-60.

45. Tang H, Lévy F, Berger H, Schmid P: Urbach tail of anatase TiO$_2$. *Phys Rev B.* 1995, 52:7771-4.

46. Mathews NR, Morales ER, Cortés-Jacome MA, Toledo Antonio JA: TiO$_2$ thin films - Influence of annealing temperature on structural, optical and photocatalytic properties. *Sol Energy.* 2009, 83:1499-508.

47. Gouttebaron R, Cornelissen D, Snyders R, Dauchot JP, Wautelet M, Hecq M: XPS study of TiOx thin films prepared by d.c. magnetron sputtering in Ar-O2 gas mixtures. Surf. *Interface Anal* 2000, 30:527-30.

48. Hashimoto S, Tanaka A: Alteration of Ti 2p XPS spectrum for titanium oxide by low-energy Ar ion bombardment. *Surf Interface Anal.* 2002, 34:262-5.

49. Olson CL, Nelson J, Islam MS: Defect chemistry, surface structures, and lithium insertion in anatase TiO_2. *J Phys Chem B.* 2006, 110:9995-10001.

50. Shi JY, Chen J, Feng ZC, Chen T, Lian YX, Wang XL, *et al.*: Photoluminescence characteristics of TiO_2 and their relationship to the photoassisted reaction of water/methanol mixture. *J Phys Chem C.* 2007, 111:693-9.

51. Mercado CC, Knorr FJ, McHale JL, Usmani SM, Ichimura AS, Saraf LV: Location of hole and electron traps on nanocrystalline anatase TiO_2. *J Phys Chem C.* 2012, 116:10796-804.

52. McHale JL, Rich CC, Knorr FJ: Trap state photoluminescence of nanocrystalline and bulk TiO_2: implications for carrier transport. *MRS Proc.* 2010, 1268:EE03-8.

53. Diebold U: The surface science of titanium dioxide. *Surf Sci Rep.* 2003, 48:53-229.

Citations

CHAPTER 1

Wei Fu, Xin-Hao Li, Hong-Liang Bao, Kai-Xue Wang, Xiao Wei, Yi-Yu Cai, and Jie-Sheng Chen, Synergistic effect of Brønsted acid and platinum on purification of automobile exhaust gases, doi: 10.1038/srep02349.

CHAPTER 2

W. Zhuang and Z. Gong, "Gel Permeation Chromatography Purification and Gas Chromatography-Mass Spectrometry Detection of Multi-Pesticide Residues in Traditional Chinese Medicine," *American Journal of Analytical Chemistry*, Vol. 3 No. 1, 2012, pp. 24-32. doi: 10.4236/ajac.2012.31005.

CHAPTER 3

R. Chandra, V. K. Vijay, and P. M. V. Subbarao, Vehicular Quality Biomethane Production from Biogas by Using an Automated Water Scrubbing System, doi.org/10.5402/2012/904167.

CHAPTER 4

N. Azmi, H. Mukhtar and K.M. Sabil, 2011. Purification of Natural Gas with High CO_2 Content by Formation of Gas Hydrates: Thermodynamic Verification. *Journal of Applied Sciences, 11: 3547-3554.*

CHAPTER 5

A. Sordi, E. P. Silva, L. F. Milanez, D. D. Lobkov, and S. N. M. Souza, Hydrogen from Biomass Gas Steam Reforming for Low Temperature Fuel Cell: Energy and Exergy Analysis, ISSN 0104-6632.

CHAPTER 6

Yang Jiang, Gizela Mikova, Robbert Kleerebezem, Luuk AM van der Wielen, and Maria C Cuellar, Feasibility Study of an Alkaline-based Chemical Treatment for the Purification of Polyhydroxybutyrate Produced by a Mixed Enriched Culture, doi:10.1186/s13568-015-0096-5.

CHAPTER 7

Sergio M. Salcedo Martínez, Guadalupe Gutiérrez-Soto, Carlos F. Rodríguez Garza, Tania J. Villarreal Galván, Juan F. Contreras Cordero and Carlos E. Hernández Luna (2013). Purification and Partial Characterization of a Thermostable Laccase from Pycnoporus sanguineus

CS-2 with Ability to Oxidize High Redox Potential Substrates and Recalcitrant Dyes, Applied Bioremediation - Active and Passive Approaches, Dr. Yogesh Patil (Ed.), ISBN: 978-953-51-1200-6, InTech, DOI: 10.5772/56374.

CHAPTER 8

Chunyan Jin, Ben Liu, Zhongxiang Lei, and Jiaming Sun, Structure and photoluminescence of the TiO_2 films grown by atomic layer deposition using tetrakis-dimethylamino titanium and ozone, doi:10.1186/s11671-015-0790-x.

Index